I0042142

ENERGY STORAGE AND CONVERSION DEVICES

Emerging Materials and Technologies
Series Editor
Boris I. Kharissov

ENERGY STORAGE AND CONVERSION DEVICES

Supercapacitors, Batteries, and Hydroelectric Cells

Edited by
Anurag Gaur, A.L. Sharma, and Anil Arya

CRC Press
Taylor & Francis Group
Boca Raton London New York

CRC Press is an imprint of the
Taylor & Francis Group, an **informa** business

First edition published 2022
by CRC Press
6000 Broken Sound Parkway NW, Suite 300, Boca Raton, FL 33487-2742

and by CRC Press
2 Park Square, Milton Park, Abingdon, Oxon, OX14 4RN

© 2022 Taylor & Francis Group, LLC

CRC Press is an imprint of Taylor & Francis Group, LLC

Reasonable efforts have been made to publish reliable data and information, but the author and publisher cannot assume responsibility for the validity of all materials or the consequences of their use. The authors and publishers have attempted to trace the copyright holders of all material reproduced in this publication and apologize to copyright holders if permission to publish in this form has not been obtained. If any copyright material has not been acknowledged please write and let us know so we may rectify it in any future reprint.

Except as permitted under U.S. Copyright Law, no part of this book may be reprinted, reproduced, transmitted, or utilized in any form by any electronic, mechanical, or other means, now known or hereafter invented, including photocopying, microfilming, and recording, or in any information storage or retrieval system, without written permission from the publishers.

For permission to photocopy or use material electronically from this work, access www.copyright.com or contact the Copyright Clearance Center, Inc. (CCC), 222 Rosewood Drive, Danvers, MA 01923, 978-750-8400. For works that are not available on CCC please contact mpkbookspermissions@tandf.co.uk

Trademark notice: Product or corporate names may be trademarks or registered trademarks and are used only for identification and explanation without intent to infringe.

Library of Congress Cataloging-in-Publication Data
Names: Gaur, Anurag, editor. | Sharma, A. L., editor. | Arya, Anil, editor.
Title: Energy storage and conversion devices : supercapacitors, batteries, and hydroelectric cells / edited by Anurag Gaur, A.L. Sharma, and Anil Arya.
Description: First edition. | Boca Raton, FL : CRC Press, 2022. |
Series: Emerging materials and technologies | Includes bibliographical
references and index.
Identifiers: LCCN 2021026424 | ISBN 9780367694258 (hbk) | ISBN 9780367694289 (pbk) |
ISBN 9781003141761 (ebk)
Subjects: LCSH: Electric batteries. | Energy storage.
Classification: LCC TK2896 .E768 2022 | DDC 621.31--dc23
LC record available at https://lccn.loc.gov/2021026424

ISBN: 978-0-367-69425-8 (hbk)
ISBN: 978-0-367-69428-9 (pbk)
ISBN: 978-1-003-14176-1 (ebk)

DOI: 10.1201/9781003141761

Typeset in Times
by MPS Limited, Dehradun

Contents

Foreword

This book presents state-of-the-art outlines of the research and development in designing the electrode and electrolyte materials for energy storage/conversion devices (Li-ion batteries and supercapacitors). Further, green energy production through water splitting by an emerging device (hydroelectric cell, HEC) is also explored. Chapters are focused on the fundamentals of the battery, supercapacitor, and HEC and deliver a synopsis of the development and selection criteria of numerous kinds of electrode and electrolyte material for developing competent devices. Diverse synthesis methods are given for the readers for attaining structural and electrochemical properties. Also, different performance parameters and their relation with structure and morphology have been explored. One chapter explores electricity generation by dissociating water through the HEC, which is a nontoxic and green source of energy production. The last chapter of the book explores challenges faced in this field and offers a vision for next-generation energy devices. This book provides a comprehensive overview of concepts and principles for three different devices and will be highly beneficial for the broad category of students/professionals in Physics, Material Science, Chemistry, and Chemical Engineering.

— **Anurag Gaur, PhD**
A. L. Sharma, PhD
Anil Arya, PhD

Preface

Energy requirements are always a top priority in any society across the globe. Continuous consumption, rapid population growth, and exhaustion of traditional resources (such as fossil fuels, coal, and biomass) encourage the scientists working in this field to develop some better alternatives. At the same time, the conventional source of energy is not appropriate to provide us the backup shortly. In the contemporary period, they are rapidly polluting the environment (global warming and air pollution), resulting in the execution of environmental norms at national/international forums. The only feasible alternative is to switch from conventional/traditional to renewable/clean green sources of energy.

The substitute renewable sources of energy are hydro energy, solar energy, wind energy, and tidal energy. These alternative sources certainly fulfill the need for energy, but are incapable of storing the energy to provide the continuous supply. To eliminate these issues of storage, the most reliable, convenient, and efficient means are batteries and supercapacitors (as they can deliver the energy as per need and can be moved from one place to another as per requirement). Another alternative renewable, eco-friendly, and green energy source is HEC because it generates power just by using a few drops of water. HEC has attracted a lot of attention globally as it has proved to be an auspicious alternative for the harvesting of green and clean energy without liberating any toxic waste.

In this book, we provided a glimpse of systematic advancement toward energy storage and conversion devices especially supercapacitors, batteries, and HECs. The design and optimization of electrode and electrolyte materials with a key focus toward augmentation of electrochemical performance have been discussed systematically. One separate chapter explores electricity generation by dissociating water through the HEC, which is a nontoxic and green source of energy production. In brief, this book offers noteworthy intuitions vis-à-vis the state-of-the-art overview of R&D in designing the electrode and electrolyte materials for Li-ion batteries, supercapacitors, and HECs. This book also explores challenges faced in this field and offers a vision for next-generation smart and efficient energy storage/conversion devices.

— **Anurag Gaur, PhD**
A. L. Sharma, PhD
Anil Arya, PhD

About the Editors

Dr Anurag Gaur, Department of Physics, National Institute of Technology, Kurukshetra, India
Dr Anurag Gaur is an Assistant Professor in the Department of Physics, National Institute of Technology, Kurukshetra. Dr Gaur has published more than 100 research articles in peer-reviewed reputed journals and has led or been involved in over 7 national and international projects funded by various government agencies (e.g. SERB-DST, CSIR, etc). He has 14 years of research experience in nanomaterials synthesis and developed supercapacitors, lithium-ion batteries, hydroelectric cells, and spintronics devices. He has guided 7 PhD and 50 M Tech students for their thesis work. His h-index is 23 and i10-index is 50 with 1500 citations and filled 3 patents. He has served as a reviewer for several high-impact journals and delivered over 50 talks at various national/international conferences. He did his PhD from the Indian Institute of Technology, Roorkee in 2007. He has been granted the Best Faculty Award by the National Institute of Technology, Kurukshetra in 2018. Dr Gaur's research interests include energy storage devices and green energy production through water splitting.

Dr A.L. Sharma, Department of Physics, Central University of Punjab, Bathinda, India
Dr A.L. Sharma is an Assistant Professor at the Department of Physics, Central University of Punjab, Bathinda. Dr Sharma has authored and co-authored over 70 publications in peer-reviewed journals including research publications, review articles, conference proceedings, and 9 book chapters in international publishers. He has 11 years of research experience and his research interests include the development of electrode and electrolyte material for lithium-ion batteries and supercapacitors. His Scopus h-index is 20 and i10-index is 27 with over 919 citations. He has served as a reviewer for several high-impact journals and given more than 20 talks at various conferences/colleges and universities. He pursued his PhD at the Department of Physics, Indian Institute of Technology, Kharagpur. Dr Sharma's research interests include the development of nanostructured materials/composites for application in energy storage/conversion devices (battery and supercapacitor).

Dr Anil Arya, Department of Physics, National Institute of Technology, Kurukshetra, India
Dr Anil Arya is serving as an Assistant Professor at the Department of Physics, National Institute of Technology, Kurukshetra. Dr Arya has co-authored 40 research publications, review articles, conference proceedings, and 9 book chapters. His Scopus h-index is 12 with over 498 citations. He has served as a reviewer for several electrochemistry-related journals. He pursued his PhD at the Department of Physics, Central University of Punjab, Bathinda. Before joining the PhD program, he received his bachelor's degree in science from university college (presently, IIHS), Kurukshetra University (Kurukshetra), and master's degree in science from the Central University of Punjab, Bathinda. Dr Arya's research interests include the synthesis of electrode/electrolyte materials for energy storage/conversion devices.

Contributors

Aarti
Department of Physics
National Institute of Technology,
    Kurukshetra
Haryana, India

Anil Arya
Department of Physics
Central University of Punjab
Bathinda, India
and
Department of Physics
National Institute of Technology,
    Kurukshetra
Haryana, India

Shamik Chakrabarti
Department of Physics
IIT Patna
Bihar, India

Anurag Gaur
Department of Physics
National Institute of Technology,
    Kurukshetra
Haryana, India

Vijay Kumar
Department of Physics
Institute of Integrated and Honors Studies
    (IIHS), Kurukshetra University
Kurukshetra, India

Lokesh Pandey
Department of Physics
Uttrakhand Technical University,
    Dehradun
Uttrakhand, India

A.L. Sharma
Department of Physics
Central University of Punjab
Bathinda, India

Meenu Sharma
Department of Mechanical Engineering
Energy Systems Research Laboratory
Indian Institute of Technology,
    Gandhinagar
Gujarat, India

Shivani Singh
Electrochemical Energy Laboratory
Department of Energy Science and
    Engineering
Indian Institute of Technology
    Bombay, Powai
Mumbai, India

1 Fundamentals of Batteries and Supercapacitors: An Overview

Anil Arya[1,2], Anurag Gaur[2], A.L. Sharma[1], and Vijay Kumar[3]
[1]Department of Physics, Central University of Punjab, Bathinda 151401, India
[2]Department of Physics, National Institute of Technology, Kurukshetra 136119, Haryana, India
[3]Department of Physics, Institute of Integrated and Honors Studies (IIHS), Kurukshetra University, Kurukshetra 136119, India

CONTENTS

DOI: 10.1201/9781003141761-1

1.1   INTRODUCTION

Energy plays an important role in daily human life and is the basic need in the present scenario. Nowadays, the development of energy sources is linked with the development of human civilization. However, the increasing demand for energy and diminution of fossil fuels has worried the scientific/industry community to look for alternative energy resources. It is a need of time to switch toward renewable sources of energy owing to the challenging issues with nonrenewable sources, e.g. increased global warming, air/water pollution, and limited stock of natural material. The top priority of the researchers is to develop sustainable and environmentally friendly sources of energy with an overall collective approach to tackle the future energy demand in portable electronics, military/space operations, household supply, electric vehicles, and power grids. The two most important energy conversion/storage devices (ECSD) are batteries and SCs that bear the potential to fulfill this increasing energy demand. The high energy density and high power density energy devices are on the radar of the research community. The SC is a very interesting candidate owing to its unique features such as the capability to deliver high instantaneous power, high power density, fast charge/discharge (in seconds), and huge cyclic stability ($>10^6$ cycles) [Choi and Aurbach 2016, Burke 2000, Simon and Gogotsi 2010, Yao et al 2015]. Table 1.1 shows the different characteristics of an SC [Wang et al 2011].

TABLE 1.1

Different Characteristics of an SC

Features	Range
Voltage level (V)	50V–100 V
Current (I)	100A–300 A
Pulse duration (Δt)	1.0 ms–1 sec
Capacitance (C)	1–10 F
Power density (KW/L)	5–180
Energy density (KJ/L)	0.5–0.6
ESR (R)	20–30
Temperature cycle	–20 °C–+60 °C

Source: Reproduced with permission from Wang et al 2011.

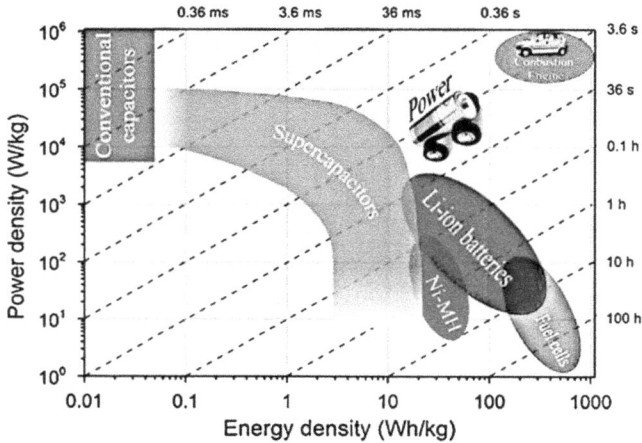

FIGURE 1.1 Ragone plots for various electrochemical energy storage systems [Reproduced with permission from Wang et al 2017].

The present chapter presents the fundamental of the SC and battery with the main focus on the classification and charge storage mechanisms. Then, important performance parameters and their interrelation are given followed by crucial characterization techniques to examine battery/SC performance. Figure 1.1 shows the Ragone plots that depict the variation in energy density against power density [Khan et al 2000, Wang et al 2015, Wang et al 2017].

It may be observed from the plot that the SC has filled the room between the batteries and the traditional capacitors owing to the high power density. Therefore, SC has gained the attention of researchers all over the globe as an alternative energy storage system for various applications. Figure 1.2a and b shows the search results for different types of battery and SCs.

1.2 PRIMARY BATTERIES

The main difference between the primary and secondary batteries is the direction of the reaction. In the former one, the electrode reactions are not reversible, whereas in the latter the electrode reactions are reversible.

1.2.1 ZINC–CARBON BATTERY (LECLANCHE CELL)

The Zn-C battery was invented by Georges-Lionel Leclanché in 1866. It comprises a zinc anode, a manganese dioxide cathode, and ammonium chloride (NH_4Cl) or zinc chloride ($ZnCl_2$) as electrolytes. As the electrode potential of zinc is –0.7 volts and the electrode potential of manganese dioxide is 1.28, therefore, the theoretical voltage of each cell is about 1.99 V. However for the commercial battery system, it is nearby 1.5 V. This battery is inexpensive and is available in different sizes. But lower energy density is one key disadvantage. The chemical reaction for NH_4Cl electrolyte is $Zn + 2MnO_2 + 2NH_4Cl + H_2O \rightarrow ZnCl_2 + Mn_2O_3 + 2NH_4OH$.

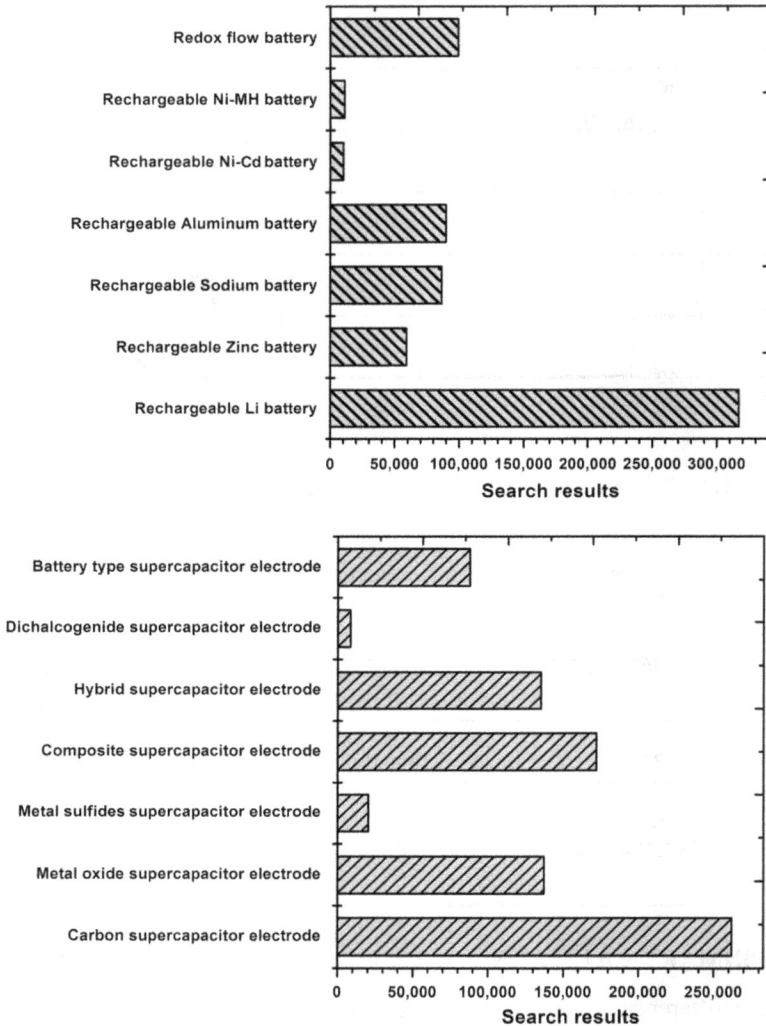

FIGURE 1.2   Search results for various keywords of (a) battery and (b) SC [Obtained on February 9, 2021, via Google Scholar].

1.2.2   ALKALINE MANGANESE BATTERY

This battery system is superior as compared to the earlier battery. Here, the reaction between the zinc and MnO_2 results in energy. The first battery was developed by Lewis Urry in 1949. In this battery, an alkaline substance (concentrated potassium hydroxide, KOH) is used as an electrolyte. The battery is capable of providing high energy density and voltage up to 1.5 V. Only disadvantage of this battery is the high cost. Reaction for this battery is $Zn + 2MnO_2 \rightleftharpoons Mn_2O_3 + ZnO$.

(a)

(b)

(c)

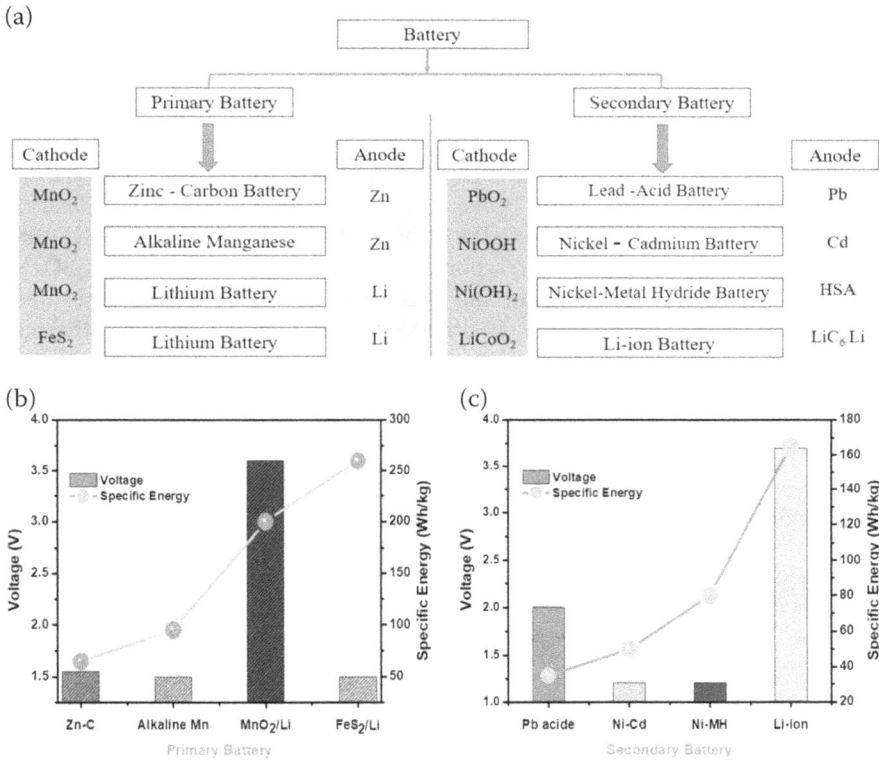

FIGURE 1.3 (a) Classification of battery systems, voltage, and specific energy comparison for (b) primary battery and (c) secondary battery.

1.2.3 LITHIUM BATTERY

Primary lithium batteries are superior in many aspects as compared to earlier types. It comprises an anode (Li foil), cathode (CuO, CuS, CF. MnO_2, etc.), and organic liquid (propylene carbonate, $LiPF_6$, and $LiClO_4$) as an electrolyte. The voltage is close to 3.5 V. The reaction is $Li + Mn^{4+}O_2 \rightarrow Mn^{3+}O_2 + Li^+$. Another alternative battery system is the lithium-iron disulfide (Li-FeS$_2$) primary battery. The unique features of this system are high capacity, low cost, low self-discharge, and low internal resistance (IR). Figure 1.3 shows the classification of battery systems, voltage, and specific energy comparison for different battery systems.

1.3 SECONDARY BATTERIES

In secondary batteries, the electrode reactions are reversible, and they can be charged and discharged via load. These batteries act as energy storage as well as a source of energy via chemical reactions. The important characteristic parameters are specific capacity, cycle life, rate capability, and safety. Important secondary battery systems are (i) lead–acid battery, (ii) nickel–cadmium battery, (iii) nickel–metal hydride battery, and (iv) secondary lithium battery (Li-ion battery).

1.3.1 LEAD–ACID BATTERY

The lead–acid battery is the first commercial rechargeable battery system invented in 1859 by French physician Gaston Planté. Here, the anode matarial is lead, and the cathode is lead dioxide (PbO_2) serves. The solution of sulfuric acid (H_2SO_4) and water is used as electrolyte. The total cell reaction can be written as $Pb + PbO_2 + 2H_2SO_4 \rightarrow 2PbSO_4 + 2H_2O$. The cell voltage for this is close to 2.1 V. These batteries are cost-effective and have a longer lifetime. But moderate energy density and low efficiency limit its use for broad applications.

1.3.2 NICKEL–CADMIUM BATTERY

Nickel–Cadmium (Ni-Cd) is the battery system invented by Ernst Waldemar Jungner in 1899 and has nickel oxide hydroxide and metallic cadmium as electrodes. Alkaline (KOH) is used as an electrolyte. The toxic nature of Cd (a heavy metal) is one disadvantage. The typical cell voltage for this is 1.2 V, and energy density is about 50–75 Wh/kg with longer cycle life. The overall net reaction is $Cd + NiO_2 + 2H_2O \rightarrow Cd(OH)_2 + Ni(OH)_2$.

1.3.3 NICKEL–METAL HYDRIDE BATTERY

Nickel–Metal Hydride is one of the most advanced commercially available rechargeable systems and is an extension of the Ni-Cd battery. These batteries are mostly used in hybrid electric vehicles. Stanford R. Ovshinsky invented and patented this battery type in 1986 and was termed as "hydrogen ion" or "protonic" battery (as it involves the transfer and "insertion" of H^+). This battery type is preferred as an environmentally friendly system due to the absence of Cd (a toxic element). The negative electrode is a hydrogen storage alloy, the positive electrode is nickel hydroxide (NiOOH), and the electrolyte is KOH.

It works on the principle of absorption, release, and transport of hydrogen within the two electrodes. Overall net reaction is $M + Ni(OH)_2 \leftrightarrow MH + NiOOH$ (Figure 1.4). The advantages of this battery type are high energy storage capability (60–100 Wh/kg), as compared to Ni-Cd, and cell voltage is up to 1.2 V. Two disadvantages that need to be eliminated are poor low-temperature capability, high rate of self-discharge (>25% per month), and these vary with temperature.

1.3.4 SECONDARY LITHIUM BATTERY

Lithium batteries are best among all secondary batteries due to high specific energy, energy density, low self-discharge rate, and high cell voltage (3.2 V). John Goodenough, Stanley Whittingham, and Akira Yoshino were awarded the Nobel Prize in Chemistry for the developing lithium-ion battery (LIB). The commercial

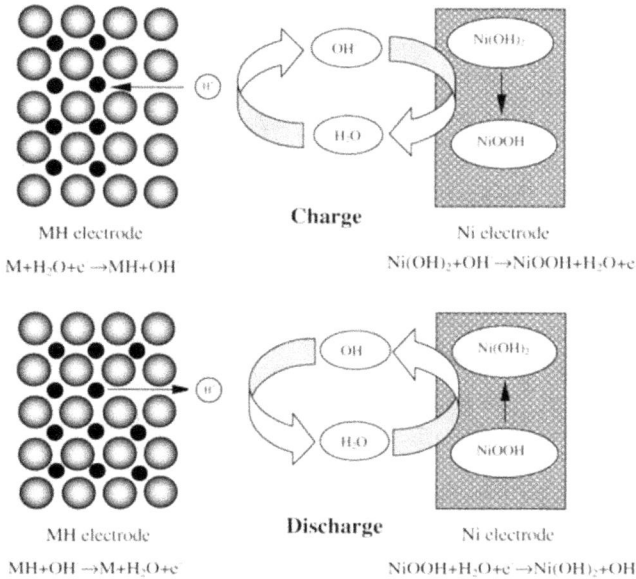

FIGURE 1.4 Schematic diagram of the electrochemical reaction process of a Ni-MH battery. For the charging process, the hydrogen atom dissociates from $Ni(OH)_2$ and is absorbed by the MH alloy. For the discharge process, the hydrogen atom dissociates from the MH alloy and combines with NiOH to form $Ni(OH)_2$ [Reproduced with permission from Feng et al 2001, Copyright Elsevier].

LIB was based on the lithium metal oxide cathode and carbon anode [Armand and Tarascon 2008]. In the charging process, lithium-ion is extracted from the cathode and moves to anode via the electrolyte. In the discharge process, lithium-ion is released from the anode toward the cathode. In both cases, the electron moves through the external circuit as the separator blocks the path of the electron (Figure 1.5).

The general battery configuration comprises three components: cathode, anode, and electrolyte-cum-separator. The cathode (generally, $LiCoO_2$) is coated on aluminum foil, and the anode (graphite) is coated on copper foil. The electrolyte is a mixture of lithium salt in organic solvents. Besides, a separator is used to separate the electrodes and is made up of polyethylene or propylene [Tarascon and Armand 2011] (Table 1.2).

Important configurations are shown in Figure 1.6i. Figure 1.6ii shows the various structures of the commercial battery system.

1.4   ELECTROCHEMICAL CHARACTERIZATION OF THE ELECTRODE MATERIAL

The electrochemical analysis of the cathode material in Li-ion batteries supports the determination of various parameters of the device like specific

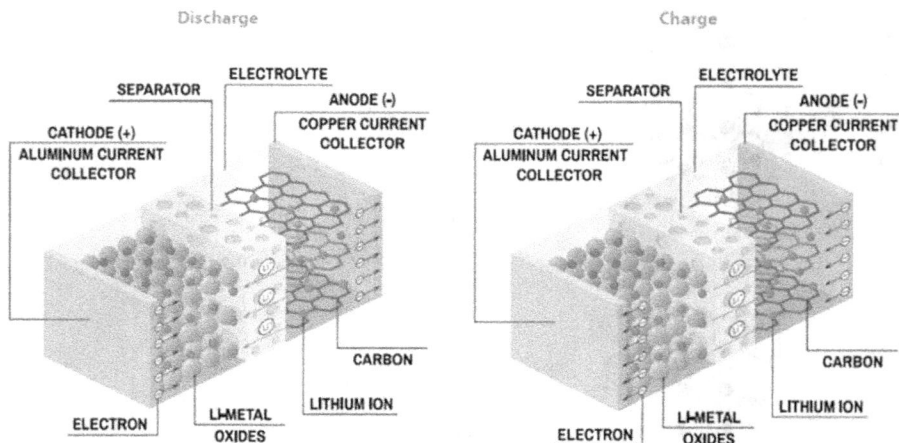

FIGURE 1.5   Discharge and charge of Li-ion battery

(Source: sivVector/Shutterstock.com).

capacitances, energy density, power density, Coulombic efficiency, and conductivity. The three-electrode and two-electrode systems are two popular systems to make electrochemical measurements. The electrochemical performances of the cathode material were analyzed via EIS (electrochemical impedance spectroscopy), CV (cyclic voltammetry), and GCD (galvanostatic charge/discharge) techniques.

1.4.1   EIS (ELECTROCHEMICAL IMPEDANCE SPECTROSCOPY)

Impedance spectroscopy is used to obtain the transport properties and dielectric properties. The real part of impedance (Z') is linked to R_s and R_{ct} as

$$Z' = R_s + R_{ct} + \sigma\omega^{-1/2} \tag{1.1}$$

Here, R_s is Ohmic resistance, R_{ct} is charge transfer resistance, and σ is the Warburg factor that is related to the diffusion coefficient of lithium-ion. The value of Ohmic resistance (R_s), charge transfer resistance (R_{ct}), and low R_{ct} value is evaluated, and low R_s and R_{ct} suggest the high specific capacity and electronic conductivity. The diffusion coefficient of Li^+ is estimated using equation 1.2

$$D = \frac{R^2 T^2}{2n^4 A^2 F^4 C^2 \sigma^2} \tag{1.2}$$

where R is the gas constant, T is the absolute temperature, A is the surface area of the cathode, n is the number of electrons per molecule during oxidization, F is the Faraday constant, and C is the concentration of lithium-ion for active electrode materials.

TABLE 1.2

Advantages and Disadvantages of Different Battery Systems

Battery type	Advantages	Disadvantages
Lead-acid	i. Mature technology ii. Worldwide production iii. Low material cost iv. No memory effect v. Low self-discharge rate vi. Relatively low capital cost	i. Short cycle life ii. Modest energy/power density iii. Long charging time iv. Safety issues (gas discharge) v. Temperature-sensitive output vi. Poor reliability
Lithium-ion (Li-ion)	i. Long cycle life ii. High round trip efficiency iii. Global R&D efforts iv. Relatively fast charging v. Highly reliable vi. Low discharge rates vii. Excellent energy/power density	i. High capital cost ii. Safety issues (thermal runaway) ii. Material bottle concerns iv. Poor recovery/recycling v. Advanced battery management systems required
Nickel–Metal hydride (NiMH)	i. Modest initial cost ii. Acceptable energy/power density iii. Modest round-trip efficiency iv. Highly reliable v. Excellent safety record vi. Relatively fast charging vii. Eco-friendly materials viii. Low operational maintenance	i. Higher self-discharge rate ii. Memory effect iii. Relatively short cycle life iv. Poor recovery/recycling
Nickel–Cadmium (Ni-Cd)	i. Comparatively low capital cost ii. Highly reliable iii. Mature technology iv. Superb safety record v. Wide operating temperatures vi. Relatively fast recharge vii. Excellent cycle life viii. Low operational maintenance	i. Modest energy/power density ii. Memory effect iii. Relatively poor round trip efficiency iv. Reliance on hazardous cadmium

Source: Reproduced with permission from Zubi et al 2018.

1.4.2 Cyclic Voltammetry (CV)

CV is a tool that is employed to investigate the electrochemistry of the cathode material. The shape of the CV curve, peak potential, and peak currents resembles the electrochemical properties of the electrode and discloses the phase transition that occurs during charge/discharge experiments, which strongly affects the capacity fading during the cycle. Mostly, when a cathode material encounters phase transformation, a peak appears in the CV curve due to the coexistence of two phases. The Li diffusion coefficients for an electrode are calculated using the

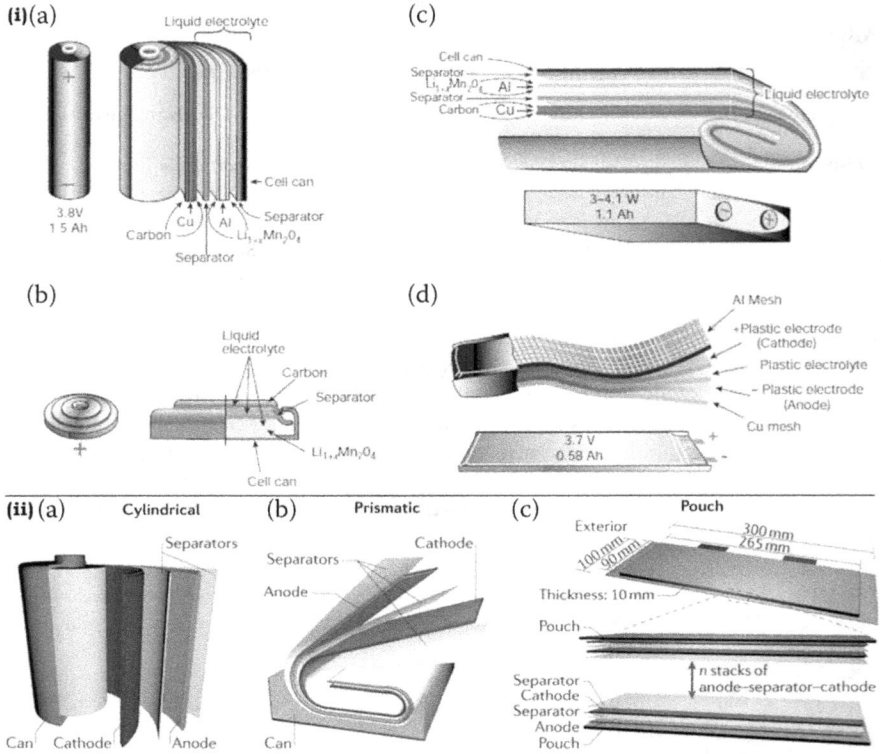

FIGURE 1.6 (i) Schematic drawing showing the shape and components of various Li-ion battery configurations. (a) Cylindrical, (b) coin, (c) prismatic, and (d) thin and flat. Note the unique flexibility of the thin and flat plastic LiIION configuration; in contrast to the other configurations, the PLiION technology does not contain free electrolyte [Reproduced with permission from Tarascon and Armand 2011]. (ii) Three representative commercial cell structures. (a) Cylindrical-type cell, (b) prismatic-type cell, and (c) pouch-type cell. The pouch dimensions are denoted, along with the internal configuration for n anode–separator–cathode stacks [Reproduced with permission from Choi and Aurbach 2016].

Randles–Sevcik equation, which describes the effect of the scan rate on the peak current. In a linear potential sweep voltammogram, the relation between the peak current and the scan rate (for low scan rates) is given by

$$i_p = 0.4463 F \left(\frac{F}{RT}\right)^{1/2} C^* S^{1/2} A D^{1/2} \tag{1.3}$$

where i_p is the peak current, F is the Faraday's constant (96,500 Cmol^{-1}), R is the gas constant (8.32 JK^{-1}mol^{-1}), T is the temperature (298.15 K), C^* is the initial Li-ion concentration for active electrode material, A is the electrode area, S is the scan rate, and D is the lithium diffusion coefficient [Hong and Zhang 2013].

TABLE 1.3
Essential Parameters for Testing the Performance of a Lithium-ion Cell

Parameters	Measuring unit	Measuring formula	Information
Operating voltage	Volts (V)	Instrumental	Energy density and safety
Current density	mAg^{-1}	Instrumental	For testing rate capabilities
Theoretical capacity	$mAhg^{-1}$	$TC = \frac{F \times x}{3.6 \times M.M \times y}$	Lithium-ion storage capability
Gravimetric capacity	$mAhg^{-1}$	$C = \frac{I(mA) \times t(h)}{m(g)}$	Li$^+$ storage capability measured per unit mass
Areal capacity	$mAhcm^{-2}$	$C = \frac{I(mA) \times t(h)}{A(cm^2)}$	Li$^+$ storage capability measured per unit area
Volumetric capacity	$mAhcm^{-2}$	$C = \frac{I(mA) \times t(h)}{V(cm^3)}$	Li$^+$ storage capability measured per unit volume
Energy density	Whg^{-1} or $Whcm^{-2}$ or $Whcm^{-3}$	$E = C \times V$	How much energy can be extracted
Power density	Wg^{-1} or Wcm^{-2} or Wcm^{-3}	$P = I \times V$	How fast the energy can be extracted
C_{rate}	h^{-1}	$C_{rate} = \frac{J(mAg^{-1})}{C(mAhg^{-1})}$	Rate of charging/discharging
Coulombic efficiency	N/A	$\%E = \frac{C_{charging}}{C_{dishcarging}} \times 100$	Reversible capacity

Source: Reproduced with permission from Gulzar et al 2016.

1.4.3 GCD (GALVANOSTATIC CHARGE/DISCHARGE)

GCD is a technique that represents the graph for the potential range vs. period for each charge/discharge cycle. The specific capacity, energy density, Coulombic efficiency, and capacity retention are estimated (Table 1.3).

1.5 SUPERCAPACITOR: AN OVERVIEW

A SC is an electrochemical device and is used to store energy and lies between the traditional capacitors and batteries. The key advantages of SC are high power density, fast charge/discharge, high-performance stability, and long cyclic stability/life (~10^6 cycles). Along with these features, only one disadvantage is the low energy density of SC. A lot of research efforts have been made to increase the energy density of SC by tuning the electrode material, electrolyte, and device structure. Figure 1.7a shows the history of SCs worldwide [Huang et al 2019]. The SC performance is influenced by the electrode material, electrolyte, and separator. These are further linked to the performance parameter of the SC cell examined by different characterization techniques. Figure 1.7b shows the schematic diagram that highlights the relation between different performance metrics, the major affecting

(a)

1996	2004	2006	2009	2010	2014	2018

Russia has developed electric cars powered by supercapacitors

Shanghai 11th has become the first commercial supercapacitor bus line in the world

German MAM develops supercapacitors for hybrid electric vehicles

Russia and Finland jointly develop flexible supercapacitors

2019

Shanghai Zhangjiang High-tech Park has completed the world's first supercapacitor bus and fast charging station system.

Nano Tecture Developed Supercapacitors for Hybrid Electric Vehicles

Shanghai Auvi Supercapacitor Bus is on its way for the first time in Sofia, Bulgaria

The United States has developed a new type of supercapacitor that can be recharged when needed

(b)

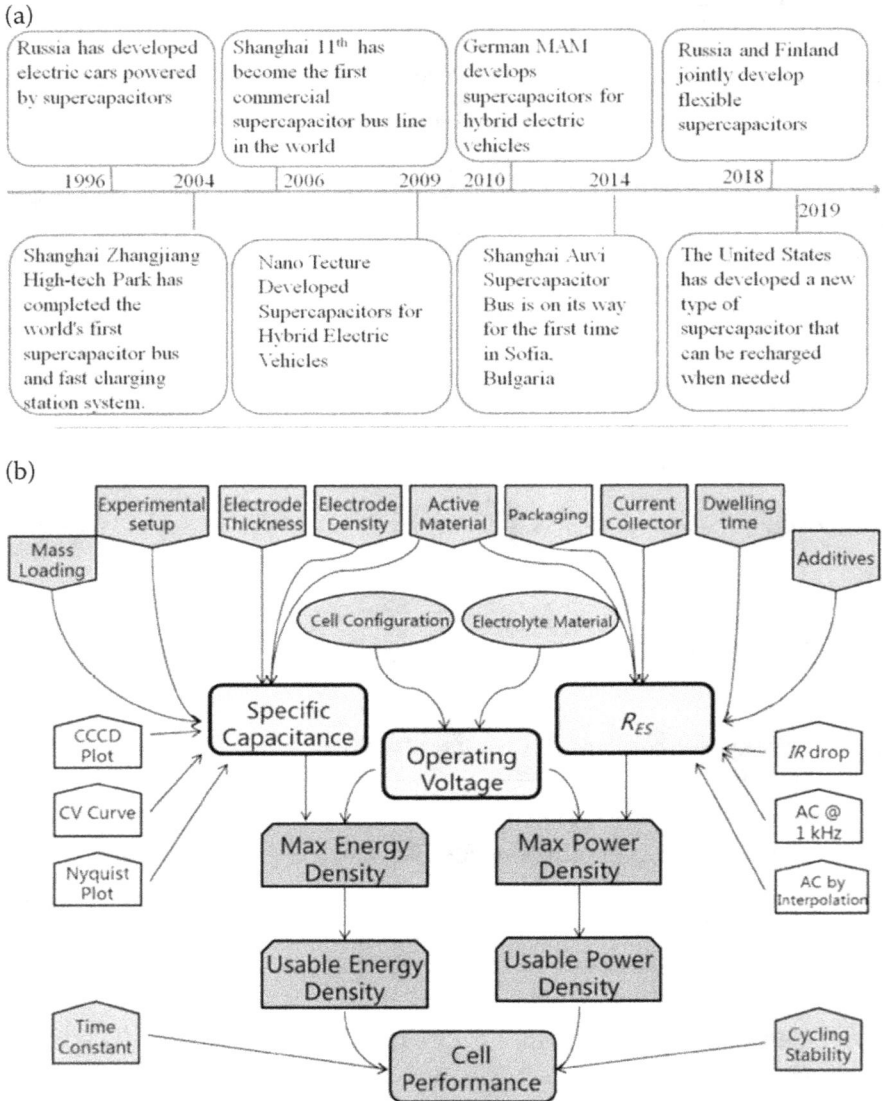

FIGURE 1.7 (a) The development of SCs in different countries [Reproduced with permission from Huang et al 2019, © AIP Publishing 2019]. (b) An illustration of key performance metrics, test methods, and major affecting factors for the evaluation of SCs [Reproduced with permission from Zhang and Pan 2015, © Wiley 2014].

factors, and the corresponding test methods. For clarity and good visibility to readers, several color schemes are employed. Three core parameters are highlighted in yellow; the power and energy densities in dark blue; time constant and cycling stability in light orange; all the important affecting factors in light purple; and the corresponding test methods in white [Zhang and Pan 2015].

1.6 FUNDAMENTALS OF SUPERCAPACITORS

SCs are also termed electrochemical capacitors owing to a similar charge storage mechanism. In simple, SC acts as a bridge between the capacitors and batteries. However, the SC stores energy by two mechanisms that are the basis of their classification: (i) electric double-layer capacitor (EDLC), energy stored via ion absorption, and (ii) pseudocapacitor (PC), energy stored by redox reactions [Salanne et al 2016, Najib and Erdem, 2019, González et al 2016, Borenstein et al 2017]. Figure 1.8 shows the classification of the SCs.

Figure 1.9 shows the comparison of the EDLC and PC with a traditional capacitor and detailed charge storage mechanism. The SC is different from the traditional capacitor or electrostatic capacitors as shown in Figure 1.9a. Depending on the charge storage mechanism, electrode material, electrolyte, and cell design are classified into three types. As SC stores energy, the charge storage phenomenon is an important criterion that decides SC performance. Based on the charge storage mechanism, SC is of three types [Meng et al 2017].

Electric double-layer capacitors (EDLCs), where the capacitance is produced by the electrostatic charge separation (no charge transport between electrode and electrolyte) at the interface between the electrode and the electrolyte (Figure 1.9b). To maximize the charge storage capacity, the electrode materials are usually made from highly porous carbon materials for achieving maximum internal surface area. The charge absorption capability is generally 0.17–0.20 electrons per atom at an accessible surface [Conw 1999, Augustyn et al 2014].

In EDLC, the ion adsorption/desorption results in charge accumulation (purely electrostatic) at the electrodes, and electrolyte ions are arranged between both electrodes (Figure 1.9b). The built-up of the high charge indicates the higher

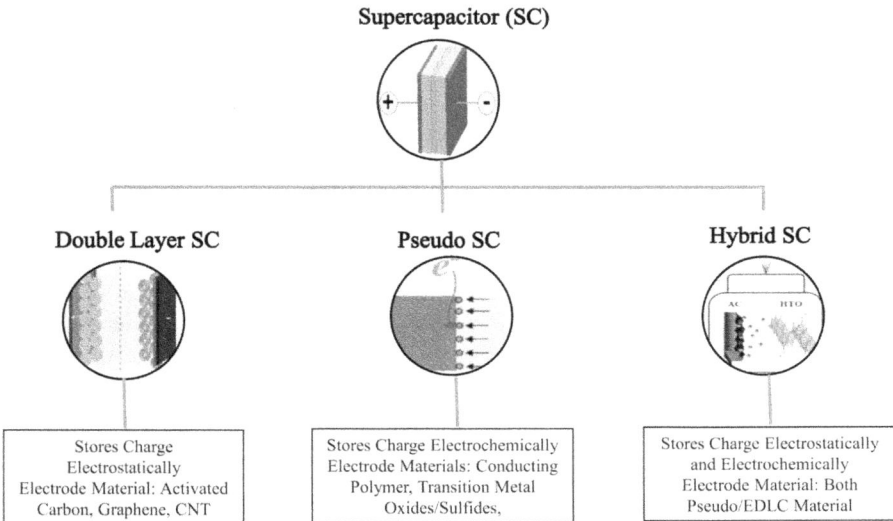

Supercapacitor (SC)

Double Layer SC — **Pseudo SC** — **Hybrid SC**

| Stores Charge Electrostatically Electrode Material: Activated Carbon, Graphene, CNT | Stores Charge Electrochemically Electrode Materials: Conducting Polymer, Transition Metal Oxides/Sulfides, | Stores Charge Electrostatically and Electrochemically Electrode Material: Both Pseudo/EDLC Material |

FIGURE 1.8 Classification of an SC.

(a)

Electrostatic capacitor

(b)

Electrical double-layer capacitor

(c)

Pseudocapacitor

(d)

Lithium ion capacitor

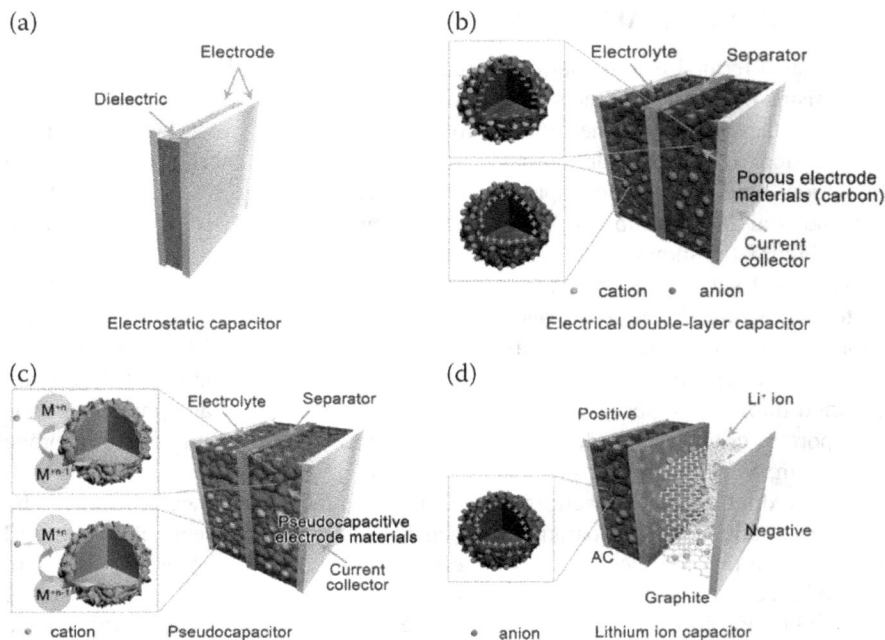

FIGURE 1.9 Schematic diagram of (a) an electrostatic capacitor, (b) an electric double-layer capacitor, (c) a pseudocapacitor, and (d) a hybrid-capacitor [Reproduced with permission from Zhong et al 2015, © Royal Society of Chemistry 2015].

capacitance. The electrodes in the case of EDLC are mostly from carbon-based material, and the absence of redox reactions is their key characteristic. The cyclic voltammetry curve is rectangular (Figure 1.10a). Pseudocapacitors (PCs) rely on fast and reversible Faradic redox reactions to store the charges at the electrode/electrolyte interface and are generally oxides/sulfides (Figure 1.10c). This is Faradic in origin and yields a charge absorption capability of ~2.5 electrons per atom at the accessible surface [Augustyn et al 2014, Sharma and Gaur 2020, Zhong et al 2015].

In PC, the charge storage is originated from the reversible redox reactions (Faradic redox reactions) taking place at the electrode/electrolyte interface (Figure 1.9c). In PC, the electrode surface area plays a very crucial role and decides the performance of the overall SC cell. The electrode material is generally a transition metal oxide. Along with the surface area, the morphology of the electrode material affects the charge storage capacity of the cell. The CV curve shows the redox peaks (Figure 1.10b–e). This also comprises the EDLC behavior and suggests higher capacitance. PC electrode materials are different from the battery electrodes as the former one does not involve any phase change [Zhong et al 2015].

Hybrid ES is a combination of two mechanisms: electrical double-layer (EDL) and Faradaic mechanisms. It is also termed as an asymmetric SC. While if one electrode material is a battery type such as PbO_2, then the device is hybrid SC. This type of SC has a high operating voltage of about 3.8 V and high specific capacitance

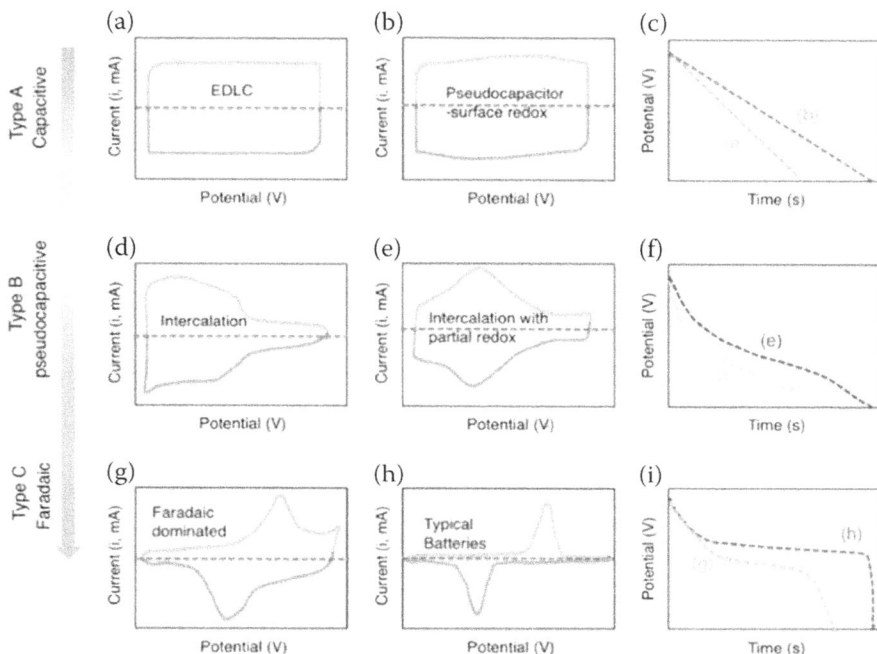

FIGURE 1.10 The CV curves of (a) EDLC materials, (b, d, e) pseudocapacitors materials, (g) battery-like materials, and (h) battery materials; and (c, f, i) corresponding galvanostatic discharge curves for various types of energy-storage materials [Reproduced with permission from Zhong et al 2015].

as well as energy density. Another important advantage is low self-discharge and low ESR. Further, based on the architecture, SC is classified as an asymmetric supercapacitor (SSC) and an asymmetric supercapacitor (ASSC). Another criterion to understand the difference between the EDLC and PC is obtained by exploring their charging/discharging and CV curves. Figure 1.10 highlights the charge/discharge and the CV curve [Gogotsi and Penner 2018]. The PC electrode material is different from the battery electrodes as evidenced in Figure 1.10.

1.7 CHARACTERISTICS OF SUPERCAPACITOR ELECTRODE MATERIAL

An SC comprises four important materials that play an important role in the SC cell performance. Therefore, before choosing the material for electrode, electrolyte, and separator, some characteristic properties need to be considered to achieve optimum performance. Since electrode material plays a very crucial role in the ion storage capacity, the high specific surface area and the favorable morphology (sheet, rod, wire, flower, wall type, etc.) of the electrode material need to be achieved initially so that the overall SC cell gives enhanced storage capacity, energy density, and power density. The electrolyte provides the ions in the SC cell, and ions migrate in

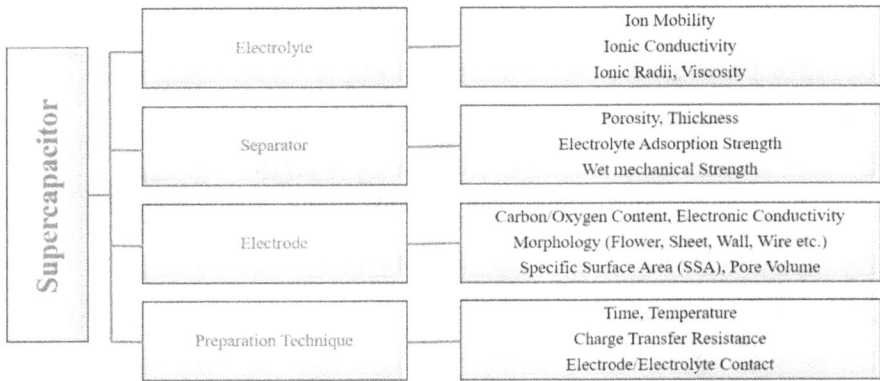

FIGURE 1.11 Crucial characteristics of different SC components

the system. Therefore, the viscosity of the electrolyte should be low, ionic radii should be small, faster ion mobility, and high ionic conductivity. Then, the separator should have high porosity to absorb the electrolyte and should have sufficient thickness. Along with the material, the preparation of the electrode material and cell influences the overall key properties of the cell. Therefore, the preparation method should be easy and cost-effective for large-scale production. The good contact of the active material on the current collector is preferred to minimize the interfacial and stability issues. Along with this during cell fabrication, the charge transfer resistance needs to be kept to a minimum to achieve maximum ion diffusion and hence the higher ion storage capacity in the electrode material. Figure 1.11 summarizes the key characteristics for the development of a high-charge storage SC.

SC has a high power density and low energy density. Hence, various strategies have been adopted by researchers to improve the energy density of the SC cell. Novel cell design (symmetric, asymmetric, and hybrid), cell voltage ($E \propto V^2$), and synthesizing new electrode nanostructures and electrolyte material open new doors of opportunity to researchers. Figure 1.12 depicts the overview of the different strategies used to improve the energy density of the SC cell [El-Kady et al 2016].

Along with important characteristics of electrode material, electrode synthesis also affects the different properties of SC. By varying the synthesis methods, the properties of the electrode can be tuned as per requirement. The important parameters are (i) electron-transfer resistance at the electrode/electrolyte interface, (ii) pseudocapacitive charging, (iii) knee frequency linked to the diffusion of electrolyte ions inside the pores of nanomaterials, (iv) resistance for ion desorption, (v) RC time constant, and (vi) dielectric relaxation time constant [Arunkumar and Paul 2017].

1.8 CHARACTERIZATION TECHNIQUES

The structural, morphological, porosity, and elemental composition properties need to be explored for the prepared electrode material before the fabrication of the SC

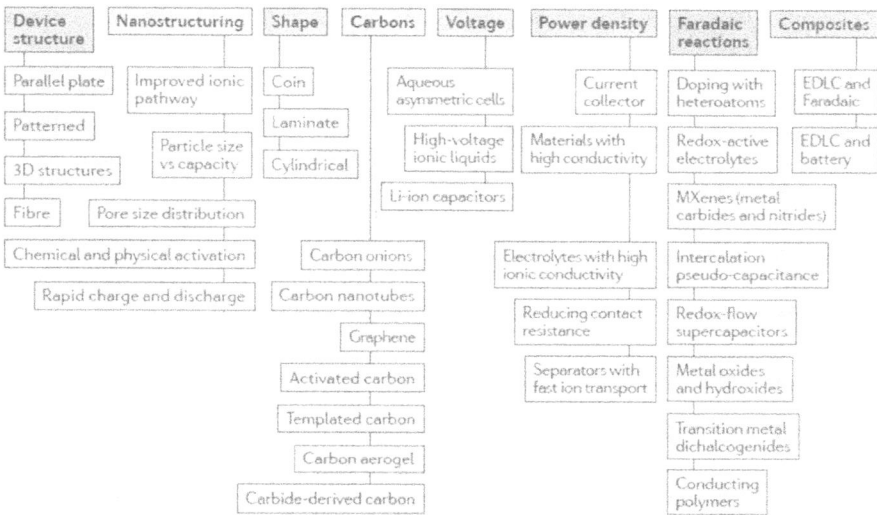

FIGURE 1.12 Strategies for improving the energy density of SCs [Reproduced with permission from El-Kady et al 2016, © Springer Nature 2016].

cell. Therefore, the important characterization techniques and information obtained from them are given below.

1.8.1 STRUCTURAL, MORPHOLOGICAL, AND SURFACE AREA STUDY

X-ray diffraction (XRD): To explore the crystallinity and form of carbon.

Raman spectroscopy: To examine the structural defects and graphitization level of carbon-based materials (by evaluating the ratio of G-band to D-band (I_G/I_D)).

Scanning electron microscopy (SEM): To explore the morphology (sheet, wire, rod, and flower) of the electrode material and composition analysis by energy dispersive spectroscopy.

Transmission electron microscopy (TEM): To investigate the thickness of the nanosheets, micropores, and mesopore diameter.

X-ray photoelectron spectroscopy (XPS): To identify the chemical nature of the C, O, and N elements.

BET-N_2 adsorption-desorption isothermal measurements: To examine the porosity, pore size/volume distribution, and the specific surface area.

1.8.2 ELECTROCHEMICAL ANALYSIS

Then, after checking the material properties and analyzing them properly, the next step is SC cell fabrication. Cell fabrication can be done in two ways: (i) symmetric cell and (ii) asymmetric cell. After that, the electrochemical properties can be analyzed in the three-electrode and two-electrode systems. The following techniques are adopted to investigate the electrochemical properties.

Complex Impedance Spectroscopy: To find out the bulk resistance (R_b), charge-transfer resistance (R_{ct}), and equivalent series resistance (ESR) of the SC cell.

The overall capacitance of the cell is $C_{Overall}^{EIS}$ (F) as obtained from impedance spectroscopy using equation 1.4

$$C_{Overall}^{EIS} = \frac{1}{2 \times \pi \times f \times Z''} \tag{1.4}$$

Here, where f is the frequency in Hz and Z'' is the imaginary part of the complex impedance in Ohm. The single electrode specific capacitance of the cell is C_{sp}^{EIS} (F/g) by multiplying the overall capacitance by a factor of 2 and divided by the mass of the active electrode material in g [Pal and Ghosh 2018].

1.8.2.1 Cyclic Voltammetry

From the CV curve, we can find out the charge storage mechanism hint by using the power law. The power law is the relation between the current and the scan rate; $i = av^b$; here, I is the cathodic current (A), v is the scan rate (mV/s), and a and b are variables [Augustyn et al 2013]. The value of b (designated as b-value) is an important factor and defines the charge storage mechanism. (i) $b = 1$ means the charge storage mechanism is a capacitive type, and (ii) $b = 0.5$ means diffusion-limited charge storage mechanism. Further, the contribution of the capacitive and diffusion capacitance can be separated using the equation: $i = k_1 v + k_2 v^{1/2}$; here, k_1 and k_2 are values [Wang et al 2007]. Here, $k_1 v$ is the capacitive contribution and $k_2 v^{1/2}$ is the diffusion-limited contribution.

1.8.2.2 Galvanostatic Charge/Discharge (GCD)

The various electrochemical parameters are obtained from the GCD using the formulas given below [Wang et al 2014].

    i. For the three-electrode system

Specific capacitance is obtained from GCD using the formula

$$C = \frac{I \times \Delta t}{m \times \Delta V} \tag{1.5}$$

Here, I is the discharging current, Δt is the discharge time, ΔV is the potential drop during discharge, and m is the mass of active material in the working electrode.

    ii. For two-electrode *(symmetric cell configuration)*

Specific Capacitance

$$C = \frac{2 \times I \times \Delta t}{m \times \Delta V} \tag{1.6}$$

Here, I is the discharging current, Δt is the discharge time, ΔV is the potential drop during discharge, and m is the mass of active material in the single electrode.

Energy Density and Power Density

$$E = \frac{1}{2 \times 3.6} C\,(\Delta V)^2 \tag{1.7}$$

$$P = \frac{3600 \times E}{\Delta t} \tag{1.8}$$

Here, E (Wh/kg), C, ΔV, P (W/kg), and Δt are the specific energy, specific capacitance, potential window, specific power, and discharge time, respectively.

   i. For two-electrode *(asymmetric cell configuration)*

Specific Capacitance

$$C = \frac{I \times \Delta t}{m \times \Delta V} \tag{1.9}$$

Here, I is the discharging current, Δt is the discharge time, ΔV is the potential drop during discharge, and m is the total mass of the active electrode materials in both (positive and negative) electrodes.

The Coulombic efficiency is calculated using the following relation:

$$\eta\,(\%) = \frac{t_d}{t_c} \times 100\% \tag{1.10}$$

Here, t_d and t_c are discharging and charging times, respectively, obtained from the charge/discharge curve [Wang et al 2014].

*   ***For Battery-Type Materials***

In hybrid devices, when battery-type materials are used as electrodes, then CV and GCD are not like EDLC and pseudocapacitor. The charge storage mechanism is like a battery, and in 2008 researchers coined the term supercapattery (supercapacitor + battery) having behavior identical to SC with high capacity and supercabattery having behavior like a rechargeable battery with high power density [Wang et al 2014, Akinwolemiwa et al 2015, Akinwolemiwa and Chen 2018, Makino et al 2012, Chen et al 2017]. Therefore, to calculate the correct electrochemical performance, the specific capacity is calculated instead of specific capacitance [Brousse et al 2015, Oyedotun et al 2019]. In such cases, a single electrode specific capacity is obtained from CV using equation

$$Q_s = \frac{1}{3.6 \times m \; S_c} \int_{E_1}^{E_2} I \times E dE \qquad (1.11)$$

where, E_1 and E_2 are the peak potentials, I (mA) is current, E (V) is the potential of an electrode, S_c (mVs^{-1}) is scan rate, and m (g) is loaded active material.

The specific capacity, Q_s (mAhg^{-1}) and energy efficiency, η_E (%) of the materials were evaluated by GCD technique using equation [Oyedotun et al 2017].

$$Q_s = \frac{I \times \Delta t}{3.6 \; m} \qquad (1.12)$$

$$\eta_E = \frac{E_d}{E_c} \times 100 \qquad (1.13)$$

Here, I (mA) is the discharge current, Δt (s) is the time taken for one discharge cycle, and m (g) is active material loading. η_E is energy efficiency, $E_c \left[= \frac{I}{3.6 \; m} \int V dt \right]$ represents the charge energy (Wh/kg), while E_d is the discharge energy obtained by integrating the area under the charge/discharge profiles, respectively. Specific capacity (C/g) is obtained via GCD using equation

$$Q = \frac{I \times \Delta t}{m} \qquad (1.14)$$

Here, I (mA) is the discharge current, Δt (s) is the discharge time, and m (g) is the mass of the active material. The optimization of charge storage performance for the positive and negative electrodes is important to achieving optimum device performance. To balance the charge ($q_- = q_+$) of both positive (q_+) and negative electrodes (q_-), the mass ratio of the positive electrode to the negative electrode in the HSC device is evaluated from CV [Zhao et al 2017].

$$q = \int \frac{im \; dV}{S_c} \qquad (1.15)$$

Here, where q (C) is the charge, i (Ag^{-1}) is the current density, m (g) is the mass of the active material, V (V) is the voltage, S_c (mVs^{-1}) as the scan rate, and $\int i dV$ is the integral area of the CV curve. And,

$$\frac{m_+}{m_-} = \frac{(i dV / S_c)_-}{(i dV / S_c)_+} \qquad (1.16)$$

Figure 1.13 shows the battery/SC electrode manufacturing process. In Figure 1.13, from left to right, the main advantages of water (as solvent) over N-Methyl-2-Pyrrolidone (NMP) are highlighted for each step. Initially, the slurry mixing is shown followed by coating on the current collector (Al for cathode, Cu for anode), the drying of the electrode layer, and, finally, the solvent recovery.

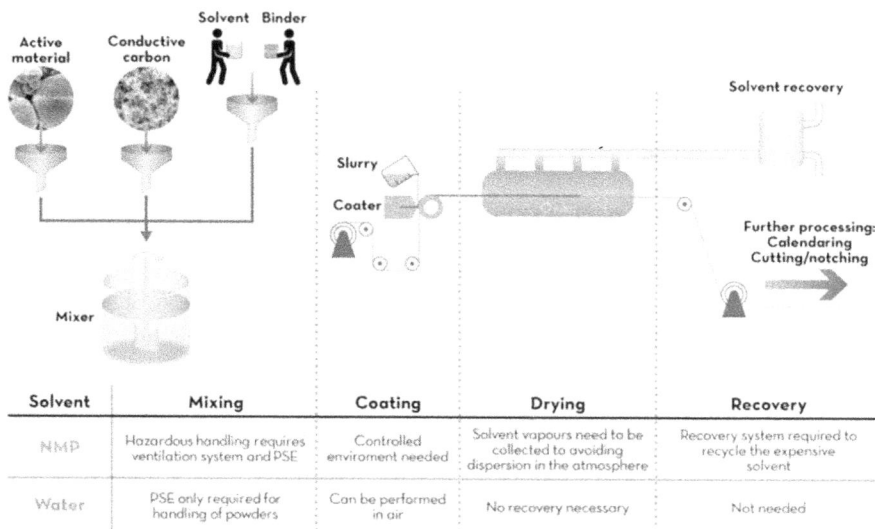

Solvent	Mixing	Coating	Drying	Recovery
NMP	Hazardous handling requires ventilation system and PSE	Controlled enviroment needed	Solvent vapours need to be collected to avoiding dispersion in the atmosphere	Recovery system required to recycle the expensive solvent
Water	PSE only required for handling of powders	Can be performed in air	No recovery necessary	Not needed

FIGURE 1.13 Simplified schematic description of a generic battery/EDLC electrode manufacturing process [Reproduced with permission from Bresser et al 2018].

1.8.2.3 Binder for Battery and Supercapacitor

A binder is, in general, a glue (a polymer) and plays a very critical role in holding the active material (AM) within the electrode (cathode and anode). It prevents the mechanical decay of the electrodes. Another role of the binder is to improve the adhesion to current collectors and must have little solubility in electrolyte for preventing degradation of the coated electrode. A binder material must be inert, and it affects the performance (charge storage and cyclic stability) of the battery and SC. An ideal binder must be inert, flexible, and stably holds AM at the current collector. The most common binder that has been used in battery and SC is polyvinylidene fluoride (PVDF). The amount of binder for electrode preparation is 5–20%. Generally, low concentration is preferred to eliminate the issue of resistance in PVDF [Kouchachvili et al 2014]. Some of the important requirements for a binder are given below:

    i. It must bind adequately the carbon to the aluminum for the long operation of the device with stable performance.
    ii. It must work with an as-small-as-possible percentage concerning the carbon to prevent the tuning of equivalent series resistance (ESR).
    iii. It must have good compatibility with the electrolyte.
    iv. Preferably, it should be water-based (a glue in which the dispersant is water, for ecological reasons).

Some binders employed in commercial SCs are styrene-butadiene rubber (SBR), polytetrafluoroethylene (PTFE), polyvinylidene difluoride (PVDF), carboxymethyl cellulose (CMC), polyacrylamide (PAM), poly(acrylic acid) (PAALi), and polyacrylic acid (PAA). Conducting polymers can be used as binders to maintain mechanical stability and enhance flexibility. Another type of binder is self-healing polymers, and the formation of reversible intermolecular bonds results in intrinsic healing of the polymer (if stretched or scratched). One unique feature of self-healing polymer binders is that they heal cracks and holes emerging in electrodes due to mechanical stress and/or volume changes during cell operation. Some examples are sodium alginate, carboxymethyl chitosan, and alginate [Saal et al 2020].

REFERENCES

Akinwolemiwa B., Peng C., & Chen G.Z. 2015. Redox electrolytes in supercapacitors. *Journal of Electrochemical Society*, 162, A5054–A5059.

Akinwolemiwa, B., & Chen, G.Z. 2018. Fundamental consideration for electro-chemicalengineering of supercapattery. *Journal of the Brazilian Chemical Society*, 29(5), 960–972.

Armand, M., & Tarascon, J.M. 2008. Building better batteries. *Nature*, 451, 652.

Arunkumar, M., & Paul, A. 2017. Importance of electrode preparation methodologies in supercapacitor applications. *ACS Omega*, 2(11), 8039.

Augustyn A., P. Simon, & B. Dunn, 2014 Pseudocapacitive oxide materials for high-rate electrochemical energy storage. *Energy & Environmental Science*, 7, 1597–1614.

Augustyn, V., Come, J., Lowe, M.A., Kim, J.W., Taberna, P.L., Tolbert, S.H.,... & Dunn, B. 2013. High-rate electrochemical energy storage through Li+ intercalation pseudocapacitance. *Nature Materials*, 12(6), 518.

Borenstein, A., Hanna, O., Attias, R., Luski, S., Brousse, T., & Aurbach, D. 2017. Carbon-based composite materials for supercapacitor electrodes: a review. *Journal of Materials Chemistry A*, 5(25), 12653–12672.

Bresser, D., Buchholz, D., Moretti, A., Varzi, A., & Passerini, S. 2018. Alternative binders for sustainable electrochemical energy storage–the transition to aqueous electrode processing and bio-derived polymers. *Energy & Environmental Science*, 11(11), 3096–3127.

Brousse, T., Bélanger, D., & Long, J.W. 2015. To be or not to be pseudocapacitive?. *Journal of The Electrochemical Society*, 162, A5185–A5189.

Burke, A. 2000. Ultracapacitors: why, how, and where is the technology. *Journal of Power Sources*, 91(1), 37–50.

Chen, H. , Yan, Z., Liu, X.Y., Guo, X.L., & Liu, Z.H. 2017. Rational design of microsphere and microcube $MnCO_3$ @MnO_2 heterostructures for supercapacitor electrodes. *Journal of Power Sources*, 353, 202–209.

Choi, J.W., & Aurbach, D. 2016. Promise and reality of post-lithium-ion batteries with high energy densities. *Nature Reviews Materials* 1, 16013.

Conw B.E. 1999. *Electrochemical Supercapacitors: Scientific Fundamentals and Technological Applications*. Kluwer Academic / Plenum, New York.

El-Kady, M.F., Shao, Y., & Kaner, R.B. 2016. Graphene for batteries, supercapacitors and beyond. *Nature Reviews Materials*, 1(7), 1–14.

Feng, F., Geng, M., & Northwood, D.O. 2001. Electrochemical behaviour of intermetallic-based metal hydrides used in Ni/metal hydride (MH) batteries: a review. *International Journal of Hydrogen Energy*, 26(7), 725–734.

Gogotsi, Y., & Penner, R.M. 2018. Energy storage in nanomaterials–capacitive, pseudocapacitive, or battery-like?. *ACS Nano*, 12(3), 2081–2083.

González, A., Goikolea, E., Barrena, J.A., & Mysyk, R. 2016. Review on supercapacitors: technologies and materials. *Renewable and Sustainable Energy Reviews*, 58, 1189–1206.

Gulzar, U., Goriparti, S., Miele, E., Li, T., Maidecchi, G., Toma, A., & Zaccaria, R.P. 2016. Next-generation textiles: from embedded supercapacitors to lithium ion batteries. *Journal of Materials Chemistry A*, 4, 16771–16800.

Hong, L., & Zhang, Z. 2013. Effect of carbon sources on the electrochemical performance of Li_2FeSiO_4 cathode materials for lithium ion batteries. *Russian Journal of Electrochemistry*, 49, 386–390.

Huang, S., Zhu, X., Sarkar, S., & Zhao, Y. 2019. Challenges and opportunities for supercapacitors. *APL Materials*, 7(10), 100901.

Khan, N., Mariun, N., Zaki, M., & Dinesh, L. (2000, September). Transient analysis of pulsed charging in supercapacitors. In 2000 TENCON Proceedings. Intelligent Systems and Technologies for the New Millennium (Cat. No. 00CH37119) (Vol. 3, pp. 193–199). IEEE

Kouchachvili, L., Maffei, N., & Entchev, E. (2014). Novel binding material for supercapacitor electrodes. *Journal of Solid State Electrochemistry*, 18(9), 2539–2547.

Makino, S., Shinohara, Y., Ban, T., Shimizu, W., Takahashi, K., Imanishi, N., & Sugimoto, W. 2012. 4 V class aqueous hybrid electrochemical capacitor with battery-like capacity. *RSC Advances*, 2(32), 12144–12147.

Meng, F., Li, Q., & Zheng, L. 2017. Flexible fiber-shaped supercapacitors: design, fabrication, and multi-functionalities. *Energy Storage Materials*, 8, 85–109.

Najib, S., & Erdem, E. 2019. Current progress achieved in novel materials for supercapacitor electrodes: mini review. *Nanoscale Advances*, 1, 2817–2827.

Oyedotun, K.O., Madito, M.J., Bello, A., Momodu, D.Y., Mirghni, A.A., & Manyala, N. 2017. Investigation of graphene oxide nanogel and carbon nanorods as electrode for electrochemical supercapacitor. *Electrochimica Acta*, 245, 268–278.

Oyedotun, K.O., Momodu, D.Y., Naguib, M., Mirghni, A.A., Masikhwa, T.M., Khaleed, A.A., Kebede, M., & Manyala, N. 2019. Electrochemical performance of two-dimensional Ti3C2-Mn3O4 nanocomposites and carbonized iron cations for hybrid supercapacitor electrodes. *Electrochimica Acta*, 301, 487–499.

Pal, P., & Ghosh, A. 2018. Highly efficient gel polymer electrolytes for all solid-state electrochemical charge storage devices. *Electrochimica Acta*, 278, 137–148.

Saal, A., Hagemann, T., & Schubert, U.S. 2020. Polymers for battery applications—active materials, membranes, and binders. *Advanced Energy Materials*, 2001984.

Salanne, M., Rotenberg, B., Naoi, K., Kaneko, K., Taberna, P.L., Grey, C.P.,... & Simon, P. 2016. Efficient storage mechanisms for building better supercapacitors. *Nature Energy*, 1(6), 16070.

Sharma, M., & Gaur, A. 2020. Designing of carbon nitride supported ZnCo 2 O 4 hybrid electrode for high-performance energy storage applications. *Scientific Reports*, 10(1), 1–9.

Simon, P., & Gogotsi, Y. 2010. Materials for electrochemical capacitors. In *Nanoscience And Technology: A Collection of Reviews from Nature Journals* (pp. 320–329).

Tarascon, J.M., & Armand, M. 2011. Issues and challenges facing rechargeable lithium batteries. *Nature*, 414, 359–367.

Wang, C., Zhou, E., He, W., Deng, X., Huang, J., Ding, M.,... & Xu, X. 2017. NiCo$_2$O$_4$-based supercapacitor nanomaterials. *Nanomaterials*, 7(2), 41.

Wang, H., Yi, H., Chen, X., & Wang, X. 2014. Asymmetric supercapacitors based on nano-architectured nickel oxide/graphene foam and hierarchical porous nitrogen-doped carbon nanotubes with ultrahigh-rate performance. *Journal of Materials Chemistry A*, 2(9), 3223–3230.

Wang, J.G., Kang, F., & Wei, B. 2015. Engineering of MnO$_2$-based nanocomposites for high-performance supercapacitors. *Progress in Materials Science*, 74, 51–124.

Wang, J., Polleux, J., Lim, J., & Dunn, B. 2007. Pseudocapacitive contributions to electrochemical energy storage in TiO_2 (anatase) nanoparticles. *The Journal of Physical Chemistry C*, 111(40), 14925–14931.

Wang, Z., Shuren, T.I.A.N., Guoming, X.I.A., & Kai, L.I. 2011. *Low Voltage Apparatus*, 7, 18–20.

Yao, F., Pham, D.T., & Lee, Y.H. 2015. Carbon-based materials for lithium-ion batteries, electrochemical capacitors, and their hybrid devices. *ChemSusChem*, 8(14), 2284–2311.

Zhang, S., & Pan, N. 2015. Supercapacitors performance evaluation. *Advanced Energy Materials*, 5(6), 1401401.

Zhao, B., Chen, D., Xiong, X., Song, B., Hu, R., Zhang, Q., Rainwater, H.B., Waller, G.H., Zhen, D., Ding, Y., Chen, Y., Qu, C., Dang, D., Wang, C.P., & Liu, M. 2017. A high-energy, long cycle-life hybrid supercapacitor based on graphene composite electrodes. *Energy Storage Materials*, 7, 32–39.

Zhong, C., Deng, Y., Hu, W., Qiao, J., Zhang, L., & Zhang, J. 2015. A review of electrolyte materials and compositions for electrochemical supercapacitors. *Chemical Society Reviews*, 44(21), 7484–7539.

Zubi, G., Dufo-López, R., Carvalho, M., & Pasaoglu, G. 2018. The lithium-ion battery: state of the art and future perspectives. *Renewable and Sustainable Energy Reviews*, 89, 292–308, doi: 10.1016/j.rser.2018.03.002

2 First Principle Study on LIB and Supercapacitor

Shamik Chakrabarti
Department of Physics, IIT Patna, Bihar 801106, India

CONTENTS

2.1 INTRODUCTION: BACKGROUND AND DRIVING FORCES

Evaluation of stability and feasibility of different electrode materials by density functional theory (DFT) simulation for Li-battery/supercapacitor applications has become an exciting area of research. This combined approach of research comprising simulation followed by experiments has been nowadays adopted rigorously and implemented on several electrodes for a priory prediction of performance enabling proper planning and execution of experiments including $LiMO_2$, $LiMPO_4$ [Fisher et al. 2008, Zhou et al. 2004], and Li_2MSiO_4 [Dompabolo et al. 2006, Wu et al. 2009] type material systems [M = Fe, Mn, Co, and Ni] and their different combinations for Li-ion battery (LIB) applications. Theoretical study (simulation) acts as a complementary tool to experimental techniques. It can guide an experimentalist by providing useful inputs regarding the material properties before experimentation. The desirable information may be obtained in many ways described summarily as follows:

DOI: 10.1201/9781003141761-2

i. Simulating structural properties enables an experimentalist to examine the feasibility of structural stability over wide-ranging conditions including changes in structures with variation in temperature, ionic motion, substitutional, and/or interstitial doping of either native or foreign atoms, and redox environment in the case of electrochemical reaction.

ii. Simulation of the density of states provides information regarding electronic structure including electronic bandgap, orbitals contributing valance and conduction bands and, hence, the conduction property, nature of bands, properties related to orbital overlap, oxidation states, magnetic properties, if any, with ion contents, etc.

iii. Identification of ground state from total energy/unit cell consideration, among several iso-structures, changes in energetics of the system due to change in ion contents in the lattice, and overall information on energetically favored stoichiometry is obtained a priori via calculations. This information becomes useful in selecting the most appropriate promising material among many others for different applications before real-scale experimentation.

iv. Modeling and evaluation of desirable properties of a conceived material, which may become efficient and useful for a particular application, can be known before its synthesis. Such an input provides a guiding path for experimentalists.

v. Comparison of simulated results with experimental measurements, however, later, determines the validity of the accuracy limit of the principles conceived and adopted during calculations. An exact correlation, well within tolerance limits, corroborates the success.

2.2 WHY DFT APPROACH?

Although quantum mechanics provides a formally exact theoretical basis for material chemistry, all the quantum mechanical methods developed for simulating material properties are associated with some approximations that can lead to errors and mismatch with actual experimental data. However, the amount of mismatch and/or the acceptability of errors depend on the method employed and the chemical question being asked. The most accurate method usually agrees very well with experimental results but unfortunately consumes an impractically large amount of computational time. The ideal theoretical approach, therefore, is to solve a many-particle Schrodinger equation directly for calculating material properties. However, this approach is limited by its complexity to small systems. On the other hand, DFT balances accuracy with computational cost and can be used to study larger molecules than is possible with other ab initio methods, namely, coupled-cluster theory (CCSD) [https://en.wikipedia.org/wiki/Coupled_cluster] or Moller–Plesset perturbation theory (MP2). Instead of searching for an exact or approximate solution of the Schrodinger equation, DFT is built around the evaluation of ground-state density and its corresponding energy functional and wave function. The pioneering work of Kohn and Sham [Kohn et al. 1965] provided the base theory of DFT built on the equivalence of ground-state density and the potential of a fictitious noninteracting system. The Kohn–Sham version of DFT is the most widely used many-body method for electronic structure calculations of atoms, molecules, solids, and solid surfaces. This preference reflects the efficiency of DFT compared to correlated wave function theories such as CCSD or MP2, even though accuracy is sacrificed. The advantage of DFT over other accurate methods based on the notion of wave functions can be best understood by considering the following: for a system comprising n electrons, its wave function would have three spatial coordinates for each electron and one more if the spin is considered, i.e. a total of $4n$ coordinates, whereas the electron density (ρ) depends only on three spatial coordinates, ρ_x, ρ_y, and ρ_z, independently of the number of electrons that constitute the system. Hence, while the complexity of wave function increases with the number of electrons, the electron density maintains the same number of variables, independently of the system size.

It should also be mentioned that in the case of a magnetic system consisting of spin polarization, the electronic density has two components: ρ^{up} and ρ^{down} corresponding to the spin up and spin down electronic charge density, respectively. The total electronic density, in this case, would be $\rho = \rho^{up} + \rho^{down}$. Hence, in this case, the many-body problem reduces to six coordinates, the up and down components of electronic charge density along three co-ordinate axis.

2.3 DENSITY FUNCTIONAL THEORY (DFT): A BRIEF REPORT

Many-body Hamiltonian for a molecular system consisting of electrons and nuclei can be given as

$$H = -\frac{\hbar^2}{2}\sum_i \frac{\nabla^2_{R_i}}{M_i} - \frac{\hbar^2}{2}\sum_i \frac{\nabla^2_{r_i}}{m_e} - \frac{1}{4\pi\varepsilon_0}\sum_{i,j} \frac{e^2 Z_i}{\left|R_i - r_j\right|} + \frac{1}{8\pi\varepsilon_0}\sum_{i\neq j} \frac{e^2}{\left|r_i - r_j\right|}$$

$$+ \frac{1}{8\pi\varepsilon_0}\sum_{i\neq j} \frac{e^2 Z_i Z_j}{\left|R_i - R_j\right|} \tag{2.1}$$

The mass of the nucleus at R_i is M_i, whereas the mass of electrons at r_i is m_e.

In this equation, several approximations can be made for a realistic system. As $M_i \gg m_e$, the first term, i.e. the kinetic energy of the nuclei can be ignored. This approximation is known as the Born-Oppenheimer approximation. This approximation assumed that nuclei are fixed in comparison to electronic motion. This assumption also leads to another conclusion that nuclear–nuclear interaction is constant, and hence, the last term of equation 2.1 can also be dropped down by shifting the reference level of the total energy.

System Hamiltonian now consists of the kinetic energy of the electron gas, the potential energy due to electron–electron interactions, and the potential energy of the electrons in the (now external) potential (V_{ext}) of the nuclei. The equation can be represented as

$$H = T + V + V_{ext} = E_e + V_{ext} \tag{2.2}$$

It is interesting to note that $T + V = E_e$ part of the Hamiltonian considers only a many-electron system, which is independent of the particular kind of many-electron system and hence is a universal term. System-specific information (which nuclei and on which positions) can only be obtained from V_{ext}.

In DFT, we consider an initial ground-state electronic density $\rho(r)$, which is calculated from the available structural information. The electronic density $\rho(r)$ can be expressed as

$$\rho(r) = N \sum \int d^3r_1 \int d^3r_2 \int \ldots\ldots\ldots \int d^3r_N \psi^*(r_1, r_2, \ldots r_i \neq r \ldots r_N) \psi(r_1, r_2, \ldots r_i$$

$$\neq r \ldots r_N) \tag{2.3}$$

where N = the total number of electrons/volume.

According to Hohenberg–Kohn theorem, ground-state energy is a unique function of ground-state electronic charge densities, i.e. there exists a one-to-one correspondence between the ground-state energy and wave function with ground-state electronic densities. Hence, the above equation can be inverted to know Ψ in terms of $\rho(r)$.

The density-functional approach, therefore, can be summarized in a mapped form as

$$\rho_0(r) \Rightarrow \Psi\left(\rho_0(r)\right) \Rightarrow E\left(\rho_0(r)\right) \text{ and other observables such as potential, etc.}$$

i.e. the knowledge of $\rho_0(r)$ implies the knowledge of the ground-state wave function *uniquely* and the potential, and hence of all other observables.

Therefore, the Kohn–Sham approach is applied to map the many-body problem of interacting electrons and nuclei to a one-electron reference system that leads to the same ground-state electronic density as the real system.

The Kohn–Sham equation can be expressed as

$$H_{ks}\varphi_i = \varepsilon_i\varphi_i \tag{2.4}$$

where, φ_i is a noninteracting single-particle wave function, and H_{ks} is a single-particle Kohn–Sham Hamiltonian.

$$H_{ks} = \frac{-\hbar^2}{2m_e}\vec{\nabla}_i^2 + \frac{e^2}{4\pi\varepsilon_0}\int\frac{\rho(\vec{r})(\vec{r}')}{|\vec{r}-\vec{r}'|}d\vec{r}' + V_{ext} + V_{xc} \tag{2.5}$$
$$\underbrace{\quad}_{T_e}\,\underbrace{\quad}_{V_e}\,\underbrace{\quad}_{V_{ne}}$$

Here, T_e = Kinetic energy of the electron V_e = Electrostatic energy of the electron due to Coulomb interaction with other electrons $V_{ext} = V_{ne}$ = Electrostatic energy of the electron due to Coulomb interaction of the nucleus V_{xc} = Exchange-correlation energy = exchange energy (V_x) + correlation energy (T_c) T_c = Correlation energy $\equiv$ K.E. of interacting electron - K.E. of noninteracting electron

That is, T_c is the excess kinetic energy that arises due to interaction with other electrons.

However, V_{xc} is not known formally, as it contains the difficult exchange and correlation contributions only. If it is assumed for a while that V_{xc} is known, then V_{eff}

can be written explicitly as $V_{eff} = V_e + V_{ne} + V_{xc}$. Hence, Kohn–Sham Hamiltonian can be represented as

$$H_{ks} = T_e + V_{eff} \tag{2.6}$$

where V_{eff} is the effective potential felt by a noninteracting electron (a *fictitious* particle) in the Kohn–Sham domain.

By solving the Kohn–Sham equation (2.3), we can achieve a single-particle wave function for i^{th} electron (a *fictitious* particle) φ_i. The DFT approach, therefore, establishes that ρ reproduced by φ_i is the ρ of the original interacting many-particle system. Hence,

$$\rho(r) = N \sum \left| \varphi_i^2 \right| \tag{2.7}$$

One should be aware of the fact that the single-particle wave functions φ_i are not the wave functions of the electrons. They describe mathematical quasiparticles without a direct physical meaning. Only the overall density of these quasiparticles is guaranteed to be equal to the true electron density. Further, V_{eff} depends on the density $\rho(r)$ that, in turn, is obtained from the ground-state wave function φ_i. This means that we are dealing with a self-consistent problem: the solutions (φ_i) determine the original equation (9.2), and the equation cannot be written and solved before its solution is known. An iterative procedure is therefore needed to come out of this paradox. Hence, the usual process lies in guessing some starting density ρ_0, and a Hamiltonian H_{ks1} is constructed with it. Next, the eigenvalue problem is solved, resulting in a set of solutions φ_{i1} from which a density ρ_1 can be derived. Most probably ρ_0 will differ from ρ_1. Now, both ρ_0 and ρ_1 are properly used (by following some mixing scheme [Pratt et al. 1952]) to construct H_{ks2}, which will yield a ρ_2, etc. The procedure can be set up in such a way in a self-consistent field (SCF) cycle that this series will converge to a density ρ_f that generates a H_{ksf} that again yields a solution : this final density is then consistent with the Hamiltonian and will be used to calculate system eigenfunctions.

2.4 THE EXCHANGE-CORRELATION ENERGY V_{XC}

An important step of DFT simulation is to approximate V_{xc} suitably for different materials. A widely used approximation called the local density approximation (LDA) that is based on the postulate that the exchange-correlation energy has the following form:

$$V_{xc}^{LDA} = \int \rho(r) \varepsilon_{xc}(\rho(r)) dr \tag{2.8}$$

The function $\varepsilon_{xc}(\rho)$ (exchange-correlation energy per unit charge at position r having charge density $\rho(r)$) for the homogeneous electron gas can be numerically known by quantum MC simulation. The postulate (2.6) implies that the exchange-correlation energy due to a particular density $\rho(r)$ could be found by dividing the material into infinitesimally small volumes with a constant density. Each such volume contributes to the total exchange-correlation energy by an amount equal to the exchange-correlation energy of an identical volume filled with a homogeneous electron gas with the understanding that it has the same overall density as the original material has in this volume. However, it is only a reasonable guess as no law of nature guarantees that the true V_{xc} is of this form. By construction, LDA is expected to perform well for systems with a slow varying density (for the metallic system). Further, it appears to be very accurate in many other (realistic) cases too. Still, LDA seems inconsistent for small molecules, and the reliability of calculation improves with the increasing size of the systems. It is also known to overly favor high-spin state structures.

An alternative approach, to improve LDA limitations, is to make the exchange-correlation contribution of every infinitesimal volume dependent not only on the local density in that volume but also on the density in the neighboring volumes. In other words, the gradient of the density will play a role. This approximation is, therefore, called the generalized gradient approximation (GGA). In general, GGA methods represent a significant improvement over the local method (LDA). GGA method also tends to give better total energy, structural energy differences, and energy barriers. GGA method also tends to expand and soften bonds compensating for the LDA tendency to over bind [Perdew et al. 1996, Anisimov et al. 1993]. Although GGA methods normally give reliable results for covalent, ionic, metallic, and hydrogen bridge bonds, the accuracy of the GGA method is still not enough for a correct description of many chemical aspects of molecules, namely, van der Waals interactions, ionization potentials, electron affinities, etc.

Although GGA performs, in general slightly better than LDA, there are some inherent disadvantages: (i) it comprises only one LDA exchange-correlation functional because there is a unique definition for ε_{xc}. However, GGA can be formulated with some freedom to incorporate the density gradient. Therefore, several versions of GGA exist, namely, the one by *Becke 86, Becke 88, Perdew 86, Perdew-Wang 91(PW91), Perdew et al. 92, Perdew, Burke,* and *Ernzerhof (PBE) 96* (in general, *PBE 96* i s used for the calculations). (ii) Moreover, in actual practice, one often fits a GGA functional with free parameters to a larger set of experimental data on atoms and molecules. Therefore, such a GGA calculation is not an ab initio calculation in the strict sense, as some experimental information is used (second disadvantage).

More recently, a new class of very promising DFT functionals based on the GGA has been developed by including additional semi-local information beyond the first-order density gradient contained in the normal GGAs. These methods, termed as meta-GGA (M-GGA), depend explicitly on higher-order density gradients or typically on the kinetic energy density, which involves derivatives of the occupied Kohn–Sham orbitals. These methods represent a significant improvement in the determination of properties such as atomization energies. However, they are technically more challenging with several difficulties in terms of numerical stability. Several M-GGA functionals have been developed, for example, *B95, KCIS, TPSS, VSXC.* However, these methods are not yet self-consistent.

2.5 HYBRID DENSITY FUNCTIONAL METHODS

It is also possible to go beyond standard GGA (or LDA) and include a percentage of or exact onsite Hartree–Fock exchange with GGA exchange-correlation functional in hybrid density functional (H-DFT) methods. However, the exact amount of Hartree–Fock exchange cannot be assigned from the first principles and therefore is fitted semi-empirically. Hybrid functionals have allowed a significant improvement over GGAs for many molecular properties. However, these types of functions are less popular than standard GGA (or LDA) due to difficulties in computing the exact exchange part. Examples of H-DFT methods include B3LYP, B3P86, B3PW91, B97-1, B97-2, B98, BH & HLYP, MPW1K, mpW3LYP, O3LYP, and X3LYP.

2.6 GGA+U APPROACH

It has been proved earlier [Zhou et al. 2004] that electronic structure and electrochemical charge transfer (Li de-intercalation voltage) for transition metal composites are better estimated in self-interaction corrected DFT (GGA+U method) calculation. In transition metal composites, there exists a strong Coulomb repulsion between spin-up and spin-down electrons in the localized d orbitals of transition metal atoms that are not taken into account by DFT-based calculation (GGA method). This discrepancy causes a self-interaction error in the calculation that underestimates the bandgap and Li de-intercalation voltage by almost 1.0 V [Zhou et al. 2004]. In the GGA+U method, Hubbard potential (U) takes into account this intrasite strong Coulomb repulsion, while other states are simulated in a plain GGA scheme as implemented in the original DFT. In this way, both the localized and delocalized orbitals are described by the same theory. In this formalism, the corrected total energy functional of a ground state system having occupation number $n_i (i = m_l, \sigma)$ [Singh et al. 2006] in a relevant localized atomic orbital $l (=d, f)$ with charge density ρ takes the form

$$E^{tot}(\rho, n_i) = E^{L(S)DA/GGA}(\rho) + E^U(n_i) - E^{dc}(n_i) \qquad (2.9)$$

where, $E^U(n_i) = U/2 \sum n_i n_j (i \neq j)$ = onsite electron–electron Coulombic repulsion term (U= amount of self-interaction correction applied to the theory); $E^{L(S)DA/GGA}$ = usual L(S)DA (local spin density approximation) or GGA energy functional; $E^{dc}(n_i)$= the double-counting term for electron–electron interaction that has already been taken care of by $E^{L(S)DA/GGA}$ term.

The simulation has been done in several steps. Experimental structural parameters are used as input to generate optimized geometric structures consistent with ground-state properties. Those optimized structural parameters were further used to simulate the electronic structure and electrochemical properties. A flow chart describing the simulation procedure is depicted below in Figure 2.1.

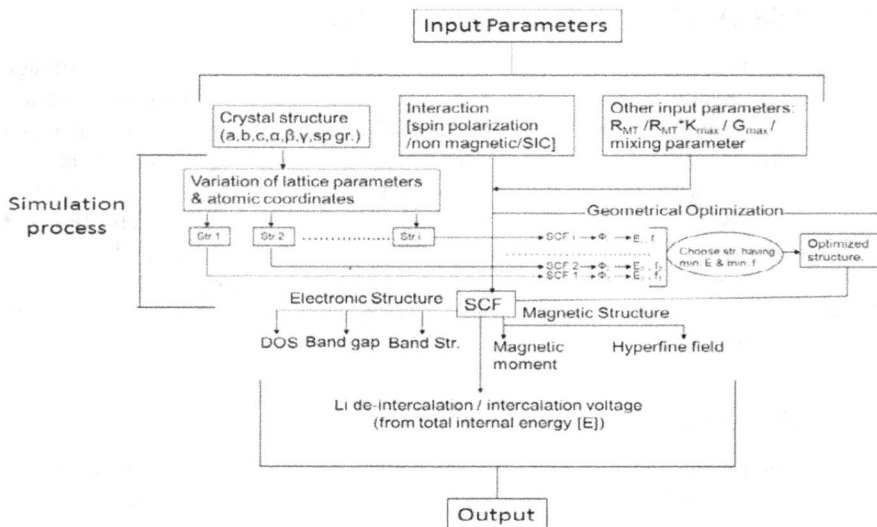

FIGURE 2.1   Flow chart describing simulation procedure adopted in DFT calculation.

2.7   STATE OF THE ART OF LI-ION BATTERY ELECTRODE MATERIALS

Solving the energy crisis for a green and safe future is a significant demand of society. To meet the requirement, several energy storage devices, namely, Li-ion Batteries (LIB), supercapacitors, and fuel cells are under intensive study. As a green source for the storage of electrochemical energy, the LIB has gained interest from its initial invention for the portable telephone by the Sony Corporation in 1991 [Ozawa et al. 1994]. They have used $LiCoO_2$ as a cathode material that could provide an electrochemical capacity of 160 mAh/g [Shukla et al. 2008] with a redox voltage of ~4 V [Shukla et al. 2008]. However, it has some disadvantages originating from the toxicity of cobalt and dendrite formation, which lead to destabilization of electrode and electrolyte materials and unavailability of capacity that could be obtained from complete Li extraction of $LiCoO_2$. An immediate alternative proposed by Padhi et al. (1997) is $LiFePO_4$. This material is exceptionally stable owing to the presence of a strong covalent P-O bond in the lattice. It exhibits a stable voltage plateau at 3.8 V [Padhi et al. 1997] with a capacity of 170 mAh/g comparable to $LiCoO_2$. However, it has a severe disadvantage that originated from its low electronic conductivity of the order of ~10^{-9} S/cm [Park et al. 2010]

Starting with layered oxides such as $LiCoO_2$, $LiNiO_2$, spinel $LiMn_2O_4$ cathode, and lithiated carbon as anode material, the development has now moved a step further aiming at high rate capability, faster redox process, and higher capacity electrodes with improved density and cycle life. To continue with the pursuits, studies on $LiMPO_4$ and Li_2MSiO_4 (where M = Fe, Co, Ni) have opened a new window for low-cost electrodes with high theoretical capacity and the possibility of intercalation by more than one Li-ion. Out of these, phosphates ($-PO_4^{-3}$) are of special interest due to their high electrochemical stability during the charge/discharge cycle arising due to strong covalent P-O bonds. On the other hand, silicates ($-SiO_4^{-4}$) are significantly important, with scope for multiple electron transfer before completion of the redox process, due to the possibility of a two-stage intercalation/de-intercalation process arising from the transport of two Li^+ ions during a redox reaction. This, in turn, doubles the electrode capacity in Li_2MSiO_4 compared to conventional electrode materials ($LiCoO_2$, $LiNiO_2$, and $LiFePO_4$) in which only one or less than one Li^+ ion per formula unit takes part in the electrochemical reaction. The search is now on for new generation of electrode materials like $LiMPO_4$, Li_2MPO_4F, $LiMBO_3$ (where M = Fe, Co, Ni), and Li_2MSiO_4-type electrode materials with better promises for capacity and energy density for LIB applications. However, phase stability with borates ($LiMBO_3$) over some time is a serious issue.

Out of these, a low-cost material phosphate (i.e. $LiFePO_4$) having an olivine structure [Petty-Weeks et al. 1988] is of special interest as it offered advantages, vis-à-vis its conventional counterpart, such as lattice stabilization due to strong P-O bond, chemical and electrochemical safety in a voltage domain, freedom from the requirements of a passivation layer, reasonably acceptable reversible capacity, etc. The only known disadvantage of $LiFePO_4$ as a cathode is the poor electronic conductivity that affects the overall power rating of the device based on it. Several approaches for an improvement of this limitation have been proposed, and the subject is under an extensive research process [Petty-Weeks et al. 1988]. Another alternative is to replace $[PO_4]^{-3}$ with $[PO_4F]^{-4}$ to form tavorite Li_2FePO_4F as new material. Li_2FePO_4F has been shown to incorporate two-dimensional Li diffusion pathways in contrast to the one-dimensional pathway of $LiFePO_4$ and comparable structural and electrochemical stability of the latter.

The ongoing quest, for a low-cost and stable cathode material with acceptable properties, has created significant interest in a wide variety of silicate structure alternatives including Li-Fe-Si-O [154-161], Li-Fe-Ti-O [Chen et al. 2015, Arillo et al. 1998, Tao et al. 2014], and Li-Fe-V-O [Liivat et al. 2010] combinations. It is expected that lower electronegativity of Si (2.03), Ti (1.54), and V(1.63) versus P (2.39) [Nyten et al. 2005] would result in (i) reduced de-intercalation voltage for Fe (II)→Fe(III) redox couple formation [Nyten et al. 2005], (ii) lowering of the electronic bandgap, and (iii) increased electronic transport. Earlier reports in the literature [Nyten et al. 2005, Nyten et al. 2006, Dominko et al. 2009, Nishimura et al. 2008, Sirisopanaporn et al. 2010, Boulineau et al. 2010, Sirisopanaporn et al. 2011, Mali et al. 2011] on this system confirmed the presence of inactive impurity as well as unreacted components (e.g. $Li_2SiO_3+FeO_x$) affecting phase purity and suitability of Li_2FeSiO_4 as an electrode material. Attempts are, therefore, going on to optimize the conditions of the synthesis process for achieving the pure phase. Further, the mechanism of voltage reduction after the first cycle is yet to be understood clearly [Nyten et al. 2006]. The problem is linked directly with the stability of the structural phase formation of Li_2FeSiO_4 and energetics related to it [Nyten et al. 2005, Nyten et al. 2006, Dominko et al. 2009, Nishimura et al. 2008, Sirisopanaporn et al. 2010, Boulineau et al. 2010, Sirisopanaporn et al. 2011, Mali et al. 2011]. However, serious attempts have been made to control the limitations on this account using the recently evolved concept of particle size engineering and novel coating strategies.

Borates ($LiMBO_3$ with M = Fe, Mn, Co, Ni) are one of the latest generations of cathode materials and are considered important for providing higher capacity (theoretical capacity 220 mAh/g for $LiFeBO_3$) originating from the low weight of $[BO_3]^{-3}$ among the all polyanion-based cathode materials. Recent studies, however, show that its performance is relatively lower than other cathode materials arising from kinetic polarization, phase stability over some time, and moisture sensitivity.

Li_2MnO_3-based cathode materials are studied [Mori et al. 2011] as one of the candidates for providing very high capacity (theoretical capacity ~459 mAh/g) and high energy density and hence could be useful in an electric vehicle. However, electrochemical inactivity of Mn^{4+} ion in this material causes poor capacity and could be improved by adding $LiMO_2$ (M = Co, Mn, Ni) [Rana et al. 2014]. Oxygen loss during the first cycle and electrolyte instability due to high voltage window (~5 V) [Rana et al. 2014] impose a limitation on $(Li_2MnO_3)_x(LiMO_2)_{(1-x)}$-type composites.

As far as anode materials for better applications are concerned, graphite or Mesocarbon microbeads (MCMB) carbon is mostly used in Li batteries. Although these materials can show high capacity (360–400 mAh/gm) and good electronic conductivity (~10^3 S/cm [Rana et al. 2014]), they have a disadvantage in terms of volume expansion/contraction during the charge/discharge cycle that causes capacity fading. Whether a Li-oxide can be used as a cathode or an anode depends on the oxidation state of the transition metals present in these composite oxides. Titanates [Reddy et al. 2013] are considered the next generation of anode materials and have also been commercialized recently. The presence of strong bonds like Ti-O makes it more electrochemically safer. A typical example of titanate-based anode material is $Li_4Ti_5O_{12}$ that shows zero strain during [Reddy et al. 2013] charge/discharge process and hence electrochemically safer than usual graphite anode. However, it has a lower capacity (165–170 mAh/g [Reddy et al. 2013]) and very low electronic conductivity (~10^{-10} S/cm at room temperature [Reddy et al. 2013]). Carbon nanotubes (CNTs) or multiwall carbon nanotubes (MWCNTs) can serve well as next-generation anode materials due to its several important characteristics, namely, (i) high electronic conductivity (~10^3–10^6 S/cm), (ii) several Li accommodation sites corresponding to CNT symmetry, which provides high intake of Li^+ ions and hence high capacity (~400–800 mAh/g), (iii) high mechanical stability with strength (~50 GPa) and shear strength (~500 MPa), (iv) good chemical stability, and (v) high active surface area. Sn-doped alloys also exhibited high capacity (~994 mAh/g) for the alloy, $Li_{4.4}Sn$. However, large volume strain (~259%) during alloying and de-alloying accompanied by the charge/discharge process acts as the main barrier in their practical applications. The solution lies in bringing the advantage of both Sn and CNT together by dispersing Sn nanoparticles within the CNT matrix. Recently, Sn-doped MWCNT has shown first discharge capacity as high as ~889 mAh/g.

2.7.1 Simulated Properties of Li-Ion Battery Electrode Materials

By DFT simulation, we can determine various properties of electrode materials from the electrochemical approach. They are described as follows:

i. **Structural stability:** The structural stability of electrode materials during charge/discharge can be studied by computing the geometric structure of both Li intercalated and Li de-intercalated structures. The XRD generated by invoking both Li intercalated and de-intercalated structures would bring a signature of structural stability during charge/discharge. If there is no change in phase apart from the change in lattice parameters, and also if the shifting of XRD peak is small indicating minute volume change, the structure would be considered as stable upon electrochemical cycling. One example is Li_2FeSiO_4 [Chakrabarti et al. 2017]. In this case, during de-lithiation, there are changes in lattice parameters while the main structural motif remains intact which leads to robust structural stability. However, if there is a change in phase due to Li intercalation or de-intercalation, there may be a large volume change that occurs during electrochemical charge/discharge, for example, SnS_2 [Hwang et al. 2018]. In SnS_2, there are four stages of lithiation: (i) Li-ion intercalation, (ii) disordering, (iii) conversion, and (iv) alloying. At the last stage of lithiation, during alloy formation, the volume expansion is >300%. The expansion in the ideal case should not exceed more than 10% during charge/discharge accompanying Li-ion intercalation/de-intercalation. In the case of $Li_4Ti_5O_{12}$, the expansion is less than 1% as it is called a zero strain material. From the structural study, we can also determine the dimension of the host lattice vacancy for guest ions, for example, Li. If the dimension is larger than the guest ion diameter, there would be reversible guest ion transfer via the host motif during charge/discharge that leads to reversible cycling.

ii. **Electronic structure:** From electronic structure, the most important properties that can be known are (i) electronic bandgap, (ii) atomic orbital responsible for electrochemical charge transfer, i.e. donor states, (iii) atomic orbital responsible for electrochemical charge acceptance, i.e. acceptor states, and (iv) atomic orbital responsible for electronic conductivity. The electronic orbital states across Fermi energy would provide the conduction electrons whereas the states at valance band maxima and conduction band minima determine donor and acceptor states, respectively. The electronic bandgap can be compared with the bandgap of other material or Li-intercalated/de-intercalated end part as obtained from pristine material to determine their conductivity qualitatively.

In the case of oxides, it may also happen that after Li de-intercalation, there are some amount of states of oxygen that crosses the Fermi energy and enters into the conduction band. In that case, this may lead to the formation of $O^{-2}/O^{-2+\gamma}$ in the reaction path and hence the possibility of oxygen evolution during charge/discharge that ultimately leads to structural instability during high voltage charging.

$Li_xABO_4 - yLi = Li_{x-y}ABO_4$ can be determined by, $E_f\left(Li_{x-y}ABO_4\right) = E\left(Li_{x-y}ABO_4\right)$
$- [E(Li_xABO_4) - E(yLi)]$.

Here, in Li_xABO_4, x amount of Li is intercalated in ABO_4 matrix; in yLi, y amount of Li is de-intercalated. $Li_{x-y}ABO_4$ is the end product obtained after y amount of Li is de-intercalated from Li_xABO_4

iii. **Identification of redox couple:** The redox couple responsible for redox reaction in the materials can be identified from the magnetic moments of constituent atoms of pristine materials and their lithiated/de-lithiated end products, for example, Li_2FeVO_4 [Chakrabarti et al. 2013]. During charging, Li_2FeVO_4 to $LiFeVO_4$ comprises a change in the magnetic moment by 0.8 μ_B for vanadium, while the magnetic moment of Fe remains intact. Hence, for this transformation, the redox couple can be identified as V^{+4}/V^{+5}. Similarly, $LiFeVO_4$ to $FeVO_4$ transformation comprises a change in the magnetic moment of Fe by 0.6 μ_B, and hence for this transformation, the redox couple can be identified as Fe^{+2}/Fe^{+3}. The presence of $O^{-2}/O^{-2+\gamma}$ redox couple can also be identified from the change in the magnetic moment of the oxygen atom during charge/discharge.

iv. **Formation energy:** The stability of the intercalated or de-intercalated end products can be determined by calculating the formation energy of the products in comparison to reactants. The formation energy E_f of the product for the chemical reaction.

If the formation energy comes out to be negative, the product will be formed. From this relation, we can determine the degree of Li de-intercalation (value of $x–y$) from pristine material.

v. **Li diffusion path and energy barrier to diffusion:** Li diffusion path can be identified from the crystal structure. Generally, the path connecting nearest-neighbor lattice sites is considered a diffusion path. $LiFeTiO_4$ consists of a Li diffusion path connecting lattice site 8a to 16c, where 8a is a crystallographic position determining tetrahedral lattice site, and 16c is a crystallographic position determining octahedral lattice site. Diffusion barrier (E_g) along diffusion path of Li can be determined by nudge elastic band method [Su et al. 2011]. The diffusion coefficient can be determined from the barrier by following the equation:

$$D = d^2.\, r.\, \exp(-E_m'K_BT) \qquad (2.10)$$

where d, r, and E_m are the hopping distance, vibrational pre-factor approximated to 1013 Hz [Su et al. 2011], and the activation barrier, respectively [Su et al. 2011]. Also, calculating diffusion barriers along different diffusion paths enables one to identify the actual path, which is the path with the lowest barrier, which will be followed by Li-ion during charge/discharge. A 3D diffusion path will lead to a higher rate capability.

Spinel compounds exhibit a diffusion path along the channel 8a→16c→8a→16c....
This path is distributed in three directions along x, y , and z. As an example, spinel
LiFeTiO4 has 3D diffusion path [Chakrabarti et al. 2016]; LiCoO2 has 2D diffusion
path whereas LiFePO4 has 1D diffusion path for the migration of Li.

vi. **Li intercalation/de-intercalation voltage:** Open circuit voltage appears due
to Li intercalation/de-intercalation for the equation $Li_{x1}MO_2 + (x2 - x1)Li =
Li_{x2}MO_2$ can be determined by an equation as proposed by Aydinol
et al. (1997).

Aydinol et al. (1997) proposed that for the voltage calculation, Gibbs free energy
(G) can be replaced by total internal energy E^{tot}. As Gibbs free energy (G) can be
represented as

$$G = E^{tot} - PV + TS \qquad (2.11)$$

Here, E^{tot} is the total internal energy, and P, V, T, and S are thermodynamic
variables denoted as pressure, volume, temperature, and entropy.

For brevity, we use the notation E in place of E^{tot} henceforth. At $T = 0$, the PV
term for solid is of the order of 10^{-5} eV and can be neglected in comparison to total
internal energy. Hence, the Li de-intercalation voltage can be evaluated as

$$V = -\frac{\Delta G}{\Delta x} = -\frac{\Delta E}{\Delta x} \qquad (2.12)$$

Here, $\Delta E = [E(Li_{x1}MO_2)-E(Li_{x2}MO_2)-(x_2-x1)E(Li)]$ and $V= -\Delta E/(x2-x1)$

2.7.2 Important Desirable Material Properties of the Electrode for Li-Ion Battery Application for Designing New Electrode Materials

Specific attributes such as high specific energy density (arising out of a right combination of working voltage and reversible capacity), high power density (owing to
acceptable electrochemical rate capability), and long cycle life (related to the stability
of structure and electrode–electrolyte interface) are simultaneously required for high-
performance storage cells. In general, structure, electrochemical stability, conductivity
(both ionic and electronic), and the available redox couples of electroactive materials
are the primary points to be considered during device design and development.
Therefore, the material components should ideally fulfill the following conditions:

i. **An open-layered structure of the cathode to permit reversible ion
migration**

In brief, the working principles of LIB devices are based on the reversible migration
of ions between cathode and anode accompanied by a redox process, whose

reversibility is a prerequisite to these devices. Thus, the structure of the electrode materials should be open-layered for easy migration of ions into its lattice. A layered structure with guest sites in the lattice facilitating ion (namely, Li^+) diffusion is required. The structure of the material should have an available Li^+ diffusion path. The diffusion path could be 1D ($LiFePO_4$), 2D (Li_2FeSiO_4), or 3D ($LiMTiO_4$ spinel). Although spinel $LiMTiO_4$ has a 3D diffusion path, migration of Li^+ becomes a bit blocked due to the presence of cationic mixing at the same lattice site.

ii. Presence of at least one transition metal element in the used material

In the cathode material, the presence of at least one transition metal element, a metal element with variable oxidation states, is required to facilitate redox reaction. Also, the presence of more than one transition metal element may activate more than one redox couple center, which would lead to the extraction of more Li^+ ions than that possible with one transition metal element and hence leads to a higher capacity. Cathode materials with multiple redox centers possessing multiple transition metal elements have higher electronic conductivity than the materials that have one transition metal element. The inclusion of multiple redox centers can be achieved by including multiple transition metal elements, such as V and Mn, in the host material, for example, $LiCoMnNiO_2$ and V- doped $LiFePO_4$ [Jiang et al. 2019]. In all these cases, the electrochemical capacity has been increased upon doping of foreign transition metal elements.

iii. Stability of electrode and electrolyte

This is the requirement for long cycle life. The insertion/extraction reaction has a topotactic character, and both insertion and extraction of guest ions into and out of the host material should ideally keep the original host structure intact. Any residual strain or hysteresis effect should be minimal over the successive charge/discharge cycle. Structural stability to sustain volume strain with ion extraction/insertion during charge/discharge requires very good mechanical strength. The presence of strong covalent bonds (e.g. P-O, Si-O, Ti-O, Sn-O, etc.) in the network structure makes an electrode safer for electrochemical reaction. Hence, during the design of new cathode materials, it would be judicious to keep these strong covalent bonds in the host structure.

iv. Higher specific energy density

The specific energy density (per weight or volume) is related to both the working voltage and the reversible capacity. The former depends on the potential of the redox process, and the latter is restricted to the reversible amount of lithium intercalation in a Li-ion battery. The available redox pair should locate in a higher and suitable potential range, and the structure of the material should be stable in a wide composition range to obtain a high capacity. High redox potential can be achieved by activating Ni^{+2}/Ni^{+4}, Fe^{+3}/Fe^{+4}, V^{+3}/V^{+4}, $O^{-2}/O^{-2+\gamma}$ redox couples in the cathode material. In general, Li-ion batteries exhibit higher energy density due to the presence of a high voltage redox couple in contrast to a supercapacitor in which redox formation voltage is relatively lower.

v. **Higher power density and electrode/electrolyte conductivity**

The electrochemical ion insertion/extraction reactions involve both diffusion of ions in the lattice and the charge transfer process on the surface of particles. Thus, the electrode's conductivity includes both ionic and electronic conductivity of the active materials. Capacity, cycle life, and rate capability mainly depend on the electronic and ionic conductivities of the electrode materials. Higher electronic conductivity is useful to keep the inner resistance low and gives an excellent power density with the first charge/discharge process in the device. Electrode materials should have higher electronic conductivity (ideally, it should be metal) and low molar mass to ensure higher capacity and rate capability.

vi. **Equivalence of mechanical and thermal coefficients of electrode and electrolyte**

During electrochemical cycles, both the electrodes undergo mechanical strain due to the extraction and insertion of ions. This also imposes strains on electrolyte materials which, if differ largely in mechanical coefficients from that of electrodes, may eventually produce strain in the electrolyte generating cracks in the device causing substantial damage. Also, the equivalence of thermal coefficients between electrodes and electrolyte provides tolerance to the device for its operation at different thermal conditions. Hence, both the electrodes and the electrolyte/separator components should have good tensile/mechanical properties ensuring volume change sustainability during cycles of operation.

viii. **Chemical insensitivity between electrode and electrolyte**

Electrode and electrolyte materials should not react with each other by forming a passivation layer, which in turn acts as an insulating barrier between ions causing the degradation of device performance ultimately. Therefore, chemical compatibility among the primary components of the device is a mandatory requirement.

ix. **A low-cost environmental benign**

The cost and environmental impact should be always kept in mind for device design. The material should be nontoxic. They are of the prime future challenges to develop low-cost and hazardless electrode/electrolyte materials with excellent device performances.

x. **Other parameters**

Longer life cycle, fast charging response, slow discharge, are performance parameters. Also, open-circuit voltage (OCV), current rating, high energy/power density, better Coulombic efficiency are key characteristic parametres of any device.

In the end, we should discuss the difficulties faced by LIB and the necessity to look for other alternatives. There are two important defects regarding LIB: (i) Li-ion batteries are expensive due to limitation of Li resources on earth's crust and

(ii) safety issues. The safety issues arise from the tendency to form a dendrite in a reaction between Li foil and anode. If the dendrites increase to reach the cathode, a short circuit occurs that eventually leads to a burst of the battery. Both of these limitations can be overcome by forming a Na-ion battery instead of LIB.

2.8   BRIEF INTRODUCTION OF SUPERCAPACITOR

Among the three main energy storage devices, namely, supercapacitor, battery, and fuel cell. Supercapacitor remains at an end with high power density. The supercapacitor is of two types: (i) EDLC and (ii) pseudocapacitor. EDLC forms capacitance by separating co-ion with counter-ion and forming a double layer at the two electrode–electrolyte interfaces. Apart from several other desirable properties, the charge storage ability depends strongly on the effective surface area of the electrode material. Pseudocapacitor forms capacitance by surface or bulk redox reaction accompanying proton (H^+) absorption in the aqueous electrolyte or ion absorption in organic/inorganic electrolyte. The mechanism of forming capacitance in EDLC is non-Faradic, while in pseudocapacitor, it is Faradic. The capacitance achieved in a pseudocapacitor is 10–100 times that of an EDLC-based supercapacitor.

2.8.1   State of the Art of Supercapacitor Electrode Materials

RuO_2 is one of the promising materials that provide specific capacitance as high as ~720 F/g [Zheng et al. 1996, Mckeown et al. 1999, Sopcic et al. 2011, Ahn et al. 2007] using its multiple redox reactions during the charge/discharge cycle. Its high cost, however, acts as a barrier to widespread practical application. This ignites the search for a low-cost electrode material having comparable structural stability and electronic conductivity. Among metal oxides, such as MnO_2 (~100–700 F/g) [Ahn et al. 2007, Wang et al. 2012, Rosario et al. 2006], iron oxides (~200 F/g) [Ahn et al. 2007, Wang et al. 2012, Rosario et al. 2006], V_2O_5 (~350 F/g) [Ahn et al. 2007, Wang et al. 2012, Rosario et al. 2006], CoO_x (~290 F/g) [Ahn et al. 2007, Wang et al. 2012, Rosario et al. 2006], and NiO_x (~250 F/g) [Ahn et al. 2007, Wang et al. 2012, Rosario et al. 2006], and conducting polymers [Wang et al. 2012] are already understudied for this purpose. Functional oxides, namely, MFe_2O_4 (M = Fe, Co, Ni) [Kuo et al. 2005] type ferrites have also been studied as electrode materials for application in supercapacitor. However, the specific capacitance of these ferrites is found to be much less (<100 F/g) [Kuo et al. 2005, Kuo et al. 2006] in comparison to other transition metal oxides due to their high molecular mass. Transition metal oxides that are capable of multiple redox reactions due to variable oxidation states of the transition metal ions or due to the presence of more than one redox center per atom are necessary for the future.

2.8.2   Simulation of EDLC and Pseudocapacitor

Modeling of EDLC requires classical density functional theory (CDFT) in contrast to LIB, where quantum DFT is needed. As modeling of EDLC requires information regarding microstructure, such as distribution of radius of pores, shape, and size of the pores, bulk structure of the capacitor, and nature of the electrolyte, a classical analog of DFT is needed where electrostatics concerning Coulombic interaction is considered. Apart from CDFT, classical molecular dynamics (CMD), quenched molecular dynamics (QMD), MC, reverse Monte Carlo (RMC), and joint DFT, etc. are used for modeling EDLC. Joint DFT utilizes self-consistent calculation for an electrode in association with the electrolyte. It comprises both quantum DFT (used to treat the electrode) and classical DFT (used to treat electrolytes). An equilibrium of electrode (solute) and electrolyte (solvent) system can be obtained by minimizing the total energy of the combined system [Zhan et al. 2017].

2.8.3 CHARACTERISTICS OF IDEAL SUPERCAPACITOR ELECTRODE MATERIAL

  i. The electrode material should have a high surface area. It will be easier to form EDLC on these high surface area electrode materials. The electrode should be formed in the nanophase to achieve a high surface area. Nanophase can be formed by using a chemical synthesis route and ball milling of the electrode material.
 ii. The material should be porous. High porosity would invoke a high surface area that is ideal for EDLC.
iii. The electrode should be metal. A metallic state invokes easier electron transfer to achieve smooth redox reactions in pseudocapacitor.
 iv. The presence of transition metal with multiple oxidation states or the presence of more than one transition metal with separate oxidation states increases the capacitance.
  v. The material should be low-cost and nontoxic.

2.9 SUMMARY

As a driving force for this work, the applications of DFT as a tool to obtain desirable information properties that can act as a guiding path for experimental realization have been documented. A brief introduction of DFT has also been addressed. DFT exhibits that the ground state wave function of a many-particle system is a unique function of ground-state electronic density, and all the material properties can be obtained from only the information of ground-state electronic density. In DFT, the many-particle equations are mapped in a fictitious noninteracting single-particle equation in which the fictitious particles "feel" the same potential as the real particle, and the density obtained from noninteracting wave functions is the same as can be obtained from the wave function of real particles. This approach is known as the Kohn–Sham approach. To solve this fictitious noninteracting single-particle equation, known as the Kohn–Sham equation, apart from kinetic energy, Coulomb repulsion (originating from electron–electron interaction) and attractive force (originating from nuclear–electron interaction of the fictitious particles), exchange-correlation energy (originated from the exchange interaction of electrons) and correlation energy (obtained from the difference of the kinetic energy of the interacting electrons and noninteracting electrons) are need to be guessed/found out. Several approximations of exchange-correlation energy are proposed, such as L(S)DA, GGA, meta-GGA, etc.

The approximations are validated by matching with experimental outputs such as bandgap, simulated XRD, voltage, etc.

A review of the current state of the art of cathode and anode materials for Li-ion batteries has been addressed. It is found that improvement of both cathode and anode materials are required to commensurate with the recent demands of energy density and rate capability for sustaining the requirement of an electric vehicle, plug-in electric vehicle, hybrid electric vehicle, or grid storage. The improvements can be made either by designing new materials or by judicious doping of foreign elements in the existing materials. DFT can serve as a tool to meet both these ends by acting as a guiding tool for the instrumentalists. DFT can be used to study structural stability of the material on Li-ion transfer during charge/discharge, the electronic bandgap of the pristine material, and its Li-ion extracted or inserted counterpart that would indicate the rate capability, redox-active ion, Li-ion diffusion path, and diffusion coefficient, and Li-ion intercalation and de-intercalation voltages, etc. The existence of intermediate states during Li-ion intercalation and de-intercalation can be known by calculating the formation energy concerning initial and final (intercalated or de-intercalated states) states.

Further, the important material properties required to be an ideal electrode material for Li-ion battery application were suggested. The electrode materials should have a layered structure with suitable guest site vacancy for Li-ion migration during charge/discharge. The presence of at least one transition metal in the electrode material is required to act as a redox-active ion. The stability of the electrode with Li-ion transfer and the compatibility of the electrolyte with the electrode is another prime issue. High energy density can be achieved via the product of high capacity and high voltage, which in turn can be, respectively, obtained by extraction/insertion of more than one Li^+ ions from the electrode material and by activating Ni^{+2}/Ni^{+4}, Fe^{+3}/Fe^{+4}, V^{+3}/V^{+4}, $O^{-2}/O^{-2+\gamma}$ redox couples in the electrode material. The electrode materials should possess a low electronic bandgap to provide high rate capability. Finally, the chosen materials should be cost-effective, nontoxic, and environmentally friendly.

A brief introduction to supercapacitor was also provided. Supercapacitors are of two types: (i) EDLC and (ii) pseudocapacitor. EDLC forms capacitance by separating positive ions with negative ions and forming a double layer at the two electrode–electrolyte interfaces, while pseudocapacitor forms capacitance by redox reaction accompanying by proton (H^+) absorption in the aqueous electrolyte or ion absorption in organic/inorganic electrolytes. Pseudocapacitors with multiple redox centers produce high specific capacitance and are desirable for future applications. A state of the art of supercapacitor indicates that there is a scope for improvement in two areas: the value of specific capacitance and cost. Simulation of EDLC requires the application of CDFT instead of quantum DFT needed for an electrode in LIB. Apart from CDFT, CMD, QMD, MC, RMC, joint DFT, etc., are used for modeling of EDLC.

Ideal supercapacitor electrode materials should be porous having high surface area, have metal with high electronic conductivity required for fast charge transfer, have multiple redox centers for getting high specific capacitance, cost-effective, nontoxic, and environmental friendly.

REFERENCES

Ahn Y.R., Park C.R., Jo S.M., Kim D.Y. 2007. Enhanced charge-discharge characteristics of RuO_2 supercapacitors on heat-treated TiO_2 nanorods. *Applied Physics Letters*, 90, 122106.

Anisimov V.I., Solovyev I.V., Korotin M.A., Czyzyk M.T., Sawatzky G.A. 1993. Density functional theory and NiO photoemission spectra. *Physical Review B*, 48, 16929–16934.

Aydinol M.K., Kohan A.F., Ceder G., Cho K., Joannopoulos J. 1997. Ab initio study of lithium intercalation in metal oxides and metal dichalcogenides. *PRB*, 56, 1354–1365.

Arillo M.A., Lopez M.L., Perez-Cappe E., Pico C., Veiga M.L. 1998. Crystal structure and electrical properties of $LiFeTiO_4$ spinel. *Solid State Ionics*, 107, 307–312.

Boulineau A., Sirisopanaporn C., Dominko R., Armstorng A.R., Bruce P.G., C. Masquelier 2010. Polymorphism and structural defects in Li_2FeSiO_4. *Dalton Transactions*, 39, 6310–6316.

Chakrabarti S., Thakur A.K., Biswas K. 2013. DFT analysis of lithiumde-intercalation in Li_2FeVO_4. *Ionics*, 19, 1515–1526.

Chakrabarti S., Thakur A.K., Biswas K. 2016. Density functional theory study of $LiFeTiO_4$. *Journal of Power Sources*, 313, 81–90.

Chakrabarti S., Thakur A.K., Biswas K. 2017. Effect of Ti modification on structural, electronic and electrochemical properties of Li_2FeSiO_4-A DFT study using FPLAPW approach. *Electrochimica Acta*, 236, 288–296.

Chen R., Knapp M., Yavuz M., Ren S., Witte R., Heinzmann R., Hahn H., Ehrenberg H., Indris S. 2015. Nanoscale spinel $LiFeTiO_4$ for intercalation pseudocapacitive Li^+ storage. *Physical Chemistry Chemical Physics*, 17, 1482–1488.

Dominko R., Arcon I., Kodre A., Hanzel D., Gaberscek M. 2009. In-situ XAS study on Li_2MnSiO_4 and Li_2FeSiO_4 cathode materials. *Journal of Power Sources*, 189, 51–58.

Dompabolo M.E.A., Armand M., Tarascon J.M., Amador U. 2006. On-demand design of polyoxianionic cathode materials based on electronegativity correlations: An exploration of the Li_2MSiO_4 system (M = Fe, Mn, Co, Ni). *Electrochemistry Communications*, 8, 1292–1298.

Fisher C.A.J., Prieto V.M.H., Islam M.S. 2008. Lithium battery materials $LiMPO_4$ (M = Mn, Fe, Co, and Ni): Insights into defect association, transport mechanisms, and doping behavior. *Chemistry of Materials*, 20, 5907–5915. https://en.wikipedia.org/wiki/Coupled_cluster.

Hwang S., Yao Z., Zhang L., Fu M., He K., Mai L., Wolverton C. 2018. SuD, multistep lithiation of tin sulfide: An investigation using in situ electron microscopy. *ACS Nano*, 12(12), 3638–3645.

Jiang S., Wang Y. 2019. Synthesis and characterization of vanadium-doped LiFePO4@C electrode with excellent rate capability for lithium-ion batteries. *Solid State Ionics*, 335, 97–102.

Kohn W., Sham L.J. 1965. Self-consistent equations including exchange and correlation effects. *Physical Review*, 140, A1133–A1138.

Kuo S.-L., Wu N.-L. 2005. Electrochemical capacitor of $MnFe_2O_4$ with NaCl electrolyte. *Electrochemical and Solid State Letters* 8(10), A495–A499.

Kuo S.-L., Wu N.-L. 2006. Electrochemical characterization on $MnFe_2O_4$/carbon black composite aqueous supercapacitors. *Journal of Power Sources*, 162, 1437–1443.

Liivat A., Thomas J.O. 2010. A DFT study of VO_4^{-3} polyanion substitution into the Li-ion battery cathode material Li_2FeSiO_4. *Computational Materials Science*, 50, 191–197.

Mali G., Sirisopanaporn C., Masquelier C., Hanzel D., Dominko R. 2011. Li_2FeSiO_4 polymorphs probed by Li MAS NMR and ^{57}Fe Mossbauer spectroscopy. *Chemistry of Materials*, 23, 2735–2744.

Mckeown D.A., Hagans P.L.,Carette L.P.L., Russell A.E., Swider K.E., Rolison D.R. 1999. Structure of hydrous ruthenium oxides: Implications for charge storage. *Journal of Physical Chemistry B*, 103, 4825–4832.

Mori D., Sakaebe H., Shikano M., Kojitani H., Tatsumi K., Inaguma Y. 2011. Synthesis, phase relation and electrical and electrochemical properties of ruthenium-substituted Li_2MnO_3 as a novel cathode material. *Journal of Power Sources*, 196, 6934–6938.

Nishimura S.I., Hayase S., Kanno R., Yashima M., Nakayama N., Yamada A. 2008. Structure of Li_2FeSiO_4. *Journal of American Chemical Society*, 130, 13212–13213.

Nyten A., Abouimrane A., Armand M., Gustafsson T., Thomas J.O. 2005. Electrochemical performance of Li_2FeSiO_4 as a new Li-battery cathode material. *Electrochemistry Communications*, 7, 156–160.

Nyten A., Kamali S., Haggstrom L., Gustafsson T.,Thomas J.O. 2006. The lithium extraction/insertion mechanism in Li_2FeSiO_4. *Journal of Material Chemistry*, 16, 2266–2272.

Ozawa K. 1994. Lithium ion rechargeable batteries $LiCoO_2$ and Carbon electrodes: the $LiCoO_2$/C system. *Solid State Ionics*, 69, 212–221, DOI: 10.1016/0167-2738(94)90411-1

Padhi A.K., Nanjundaswamy K.S., Goodenough J.B. 1997. Phospho-olivines as positive-electrode materials for rechargeable Lithium batteries. *Journal of Electrochemical Society*, 144(4), 1188–1194.

Park M., Zhang X., Chung M., Less G.B., Sastry A.M. 2010. A review of conduction phenomena in Li-ion batteries. *Journal of Power Sources*, 195, 7904–7929. DOI: 10.1 016/j.jpowsour.2010.06.060.

Perdew J.P., Burke K., Ernzerhof M. 1996. Generalized gradient approximation made simple. *Physical Review Letter*, 77, 3865–3868.

Petty-Weeks S., Zupancic J.J., Swedo J.R. 1988. Proton conducting interpenetrating polymer networks. *Solid State Ionics*, 31, 117–125.

Pratt G.W. 1952. Wave functions and energy levels for Cu^+ as found by the Slater Approximation to the Hartree-Fock Equations. *Physical Review*, 88, 1217–1224.

Rana J., Kloepsch R., Li J., Scherb T., Schumacher G., Winter M., Banhart J. 2014. On the structural integrity and electrochemical activity of $0.5Li_2MnO_3.0.5LiCoO_2$ cathode material for lithium-ion batteries. *Journal of Materials Chemistry A*, 2, 9099–9110.

Reddy M.V., Subba Rao G.V., Chowdari B.V.R. 2013. Metal oxides and oxysalts as anode materials for Li ion batteries. *Chemical Review*, 113, 5364–5457.

Rosario A.V., Bulhoes L.O.S., Pereira E.C. 2006. Investigation of pseudocapacitive properties of RuO_2 film electrodes prepared by polymeric precursor method. *Journal of Power Sources*, 158, 795–800.

Shukla A.K., Kumar T.P. 2008. Materials for next generation lithium batteries. *Current Science*, 94, 314–331.

Singh D.J., Nordstorm L. 2006. *Planewaves, Pseudopotentials and LAPW Method,* Springer, USA.

Sirisopanaporn C., Boulineau A., Hanzel D., Dominko R., Budic B., Armstorng A.R., Bruce P.G., Masquelier C. 2010. Crystal structure of a new polymorph of Li_2FeSiO_4. *Inorganic Chemistry*, 49, 7446–7451.

Sirisopanaporn C., Masquelier C., Bruce P.G., Armstorng A.R., Dominko R. 2011. Dependence of Li_2FeSiO_4 electrochemistry on structure. *Journal of American Chemical Society*, 133, 1263–1265.

Sopcic S., Rokovic M.K., Mandic Z., Roka A., Inzelt G. 2011. Mass changes accompanying the pseudocapacitance of hydrous RuO_2 under different experimental conditions. *Electrochemica Acta* 56, 3543–3548.

Su D., Ahn H., Wang, G. 2011. Ab initio calculations on Li-ion migration in Li_2FeSiO_4 cathode material with a P21 symmetry structure. *Applied Physics Letters*, 99, 141909-1–141909-3.

Tao T., Rahman M.M., Ramireddy T., Sunarso J., Chen Y., Glushenkov A.M. 2014. Preparation of composite electrodes with carbon nanotubes for lithium-ion batteries by low-energy ball milling. *RSC Advances*, 4, 36649–36655.

Wang G., Zhang L., Zhang J. 2012. A review of electrode materials for electrochemical supercapacitors. *Chemical Society Review*, 41, 797–828.

Wu S.Q., ZhuZ. Z.,Yang Y., Hou Z.F., 2009. Structural stabilities, electronic structures and lithium deintercalation in Li_xMSiO_4 (M=Mn, Fe, Co, Ni): A GGA and GGA+ U study. *Computational Material Science*, 44, 1243–1251.

Zhan C., Lian C., Zhang Y., Thompson M.W., Xie Y., Wu J., Kent P.R.C., Cummings P.T., Jiang D., Wesolowski D.J., 2017. Computational insights into materials and interfaces for capacitive energy Storage. *Advanced Science*, 4, 1700059, DOI:10.1002/advs.201700059

Zheng J.P., Jow T.R. 1996. High energy and high power density electrochemical capacitors. *Journal of Power Source*, 62, 155–159.

Zhou F., Cococcioni M., Kang K., Ceder G. 2004a. The Li intercalation potential of $LiMPO_4$ and $LiMSiO_4$ olivines with M = Fe, Mn, Co, Ni. *Electrochemistry Communications*, 6, 1144–1148.

Zhou F., Cococcioni M., Marianetti C.A., Morgan D., Ceder G. 2004b. First principles prediction of redox potentials in transition metal compounds with LDA+U. *Physical Review B*, 70, 235121-1–235121-8.

3 Cathode Materials for Li-Ion Batteries

Shivani Singh
Electrochemical Energy Laboratory, Department of Energy
Science and Engineering, Indian Institute of Technology
Bombay, Powai, Mumbai 400076, India

CONTENTS

3.1   INTRODUCTION

Oil shocks of the 1970s saw prices spiraling out of control that led the industry to think seriously in new research directions, which laid the foundation of lithium-ion batteries (LIBs). LIBs have been in use for quite some time in portable electronic devices. However, it is high time that they should move up in the hierarchy and present themselves as an attractive alternative for high scale energy applications such as electric vehicles [Dunn et al. 2011, Tarascon et al. 2010]. To make LIBs greener and a sustainable source of energy, research is directed to increase the energy density, lower its cost, and improve safety, which are few fundamental challenges [Dunn et al. 2011, Tarascon et al. 2010, Manthiram 2011]. The energy density of batteries has only increased by a factor of five over the last two centuries [Tarascon et al. 2010]. The cathode of a Li-ion battery plays a pivotal role in increasing energy density and life cycle. As reported by Tarascon et al. that to improve the energy density of a battery by 57%, cathode capacity needs to be doubled. However, an anode capacity needs to be enhanced by 10 times to increase energy density by 47% [Tarascon et al. 2010]. Hence, it is quite

DOI: 10.1201/9781003141761-3

obvious that the key to achieving a high energy density battery lies in the choice of promising cathode material that could display either greater redox potential (e.g. highly oxidizing) or higher capacity (materials capable of reversibly inserting more than one electron per transition metal) [Tarascon et al. 2010, Manthiram 2011]. This chapter will focus on a description of important properties and criteria for the selection of cathode materials. Additionally, crystal structure and electronic properties along with synthesis techniques have also been discussed.

3.2 ESSENTIAL CHARACTERISTICS OF AN EFFICIENT CATHODE MATERIAL

[Whittingham 2004, Pasquali 2004], listed below as:

I. The intercalation compound should have high lithium chemical potential to maximize the cell voltage. This implies that the transition metal ion should have a high oxidation state.

II. The material should support mixed conduction. It should have good electronic and lithium-ion conductivity to minimize polarization losses during the charge/discharge process, and thereby support high current and power density.

III. The material should reversibly react with lithium and should have good structural stability. This means that the intercalation and de-intercalation process should be reversible with no or minimal changes in the host structure to provide a good cycle life for the cell.

IV. The material should allow intercalation and de-intercalation of large amounts of lithium to maximize cell capacity. This depends on the number of available lithium sites and the accessibility of multiple valences for transition metal in the intercalation host. The energy density can be maximized by a combination of higher capacity and cell voltage as the product of voltage and capacity gives energy density.

V. The material reacts with lithium very rapidly both on intercalation and de-intercalation. This leads to high power density, which is an important characteristic to replace the existing technology of electric vehicle batteries.

VI. The material should be chemically stable without undergoing any reaction with the electrolyte over the entire range of lithium intercalation and de-intercalation.

VII. The material should be of low cost, environmentally benign, and lightweight.

VIII. The transition metal atom and Li^+ ion should not be in the same plane to avoid blockage of Li^+ ion diffusion by transition metal atom.

3.3 THREE MAJOR CLASS OF CATHODE MATERIALS: AN OVERVIEW

In the 1970s, intercalation of ions in layered dichalcogenides had been discovered. Titanium disulfide (TiS_2) was the most suitable and widely studied electrode out of all layered dichalcogenides because it has a good metallic character and undergoes lithium intercalation reversibly. Li_xTiS_2/Li cell was commercialized by Exxon and has been restricted to button size because of safety issues such as reversible deposition over lithium electrode [Nitta et al. 2015, Whittingham 2004]. $LiCoO_2$, having a similar structure to dichalcogenides, was recognized by John B. Goodenough. Electrochemical removal of lithium from $LiCoO_2$ demonstrates it as a viable cathode material [Goodenough and Kim 2010, Manthiram 2004]. Later, $LiCoO_2$ cathode along with graphite anode was demonstrated as a successful Li-ion battery by Sony. The use of graphite makes Li-ion battery much safer than lithium battery because of the substantial decrease in the dendrite formation [Nitta et al. 2015, Nazar 2004]. The cathode material that is most commonly used in commercialized LIBs is $LiCoO_2$ [Goodenough and Kim 2010]. $LiCoO_2$ is a successful cathode material, but it is expensive due to the presence of cobalt. Moreover, its toxicity leads to an increase in overall cost as proper safety measures need to be adopted for its disposal and treatment. $LiCoO_2$ can undergo performance degradation or failure when overcharged [Manthiram 2011, Nitta et al. 2015]. Other transitional metals such as Mn, Ni, and Fe are far abundantly available as well as do not show any harmful toxic behavior. $LiNiO_2$ is lower in cost and has a higher energy density (15% higher by volume and 20% higher by weight), but it is less stable and ordered compared to $LiCoO_2$ [Manthiram 2011, Nitta et al. 2015]. The lower degree of ordering results in Ni-ion occupying sites in Li planes. The addition of Co to $LiNiO_2$ increases the degree of order that leads to Ni-ion occupying sites in the Ni/Co plane rather than in the Li plane. $LiFeO_2$ is not a practical cathode as Fe migration from octahedral sites to tetrahedral sites leads to structural instability. $LiMnO_2$ is an inexpensive and environment-friendly material but can change its structure during the cycling process. The addition of Ni and Co to $LiMnO_2$ can lead to the formation of a stable structure. The mixed oxide i.e. $Li(Ni_{1/3}Mn_{1/3}Co_{1/3})O_2$ (NMC) shows high capacity, good rate capability, and can operate at high voltages. NMC cathode material can be used in electric vehicle applications due to its high power and structural stability [Ohzuku and Makimura 2001].

Another major class of cathode material is spinel manganese oxide $(LiMn_2O_4)$. $LiMn_2O_4$ is a low-cost material and much safer than $LiCoO_2$, but it has a lower capacity as compared to the layered cathode material [Xia et al. 2012]. The insertion and extraction of lithium in the case of $LiMn_2O_4$ occur in two steps. The initial cubic spinel symmetry is maintained when lithium is extracted from 8a tetrahedral sites, which occurs at around 4.0 V whereas lithium extraction/insertion from 16c octahedral sites occurs at around 3.0 V that results in cubic to tetragonal transition. The cubic to tetragonal transition results in an increase in the c/a ratio by 16%, which disturbs the structural integrity during the charge/discharge cycle leading to rapid capacity fade in the 3 V region. Therefore, limited capacity can be achieved with this material in the 4 V region; it shows a rapid loss of capacity in this region as well. Several factors such as the Jahn–Teller distortion, loss of crystallinity, and dissolution of Mn in electrolyte result in capacity fading [Manthiram 2011, Nitta et al. 2015, Xia

1997]. Several strategies have been undertaken to overcome the capacity fading such as (i) cationic substitutions with Cr, Co, and Ni can suppress Jahn-Teller distortion, and (ii) modification or coating of the surface with Al_2O_3, MgO, and V_2O_5 can suppress the dissolution of Mn in electrolyte [Manthiram 2004]. An optimum cationic substitution is found to maintain crystallinity and can also be effective in improving capacity retention during cycling.

Lithium iron phosphate ($LiFePO_4$) emerges as a promising cathode material in another class of polyanionic cathode materials. Its properties such as high capacity, i.e 160 $mAhg^{-1}$, low cost, and safety make it suitable for hybrid electric vehicles [Yamada et al. 2001, Yuan et al. 2011]. The major disadvantage with this material is the relatively low electronic conductivity, i.e. 10^{-9} S/cm for pure $LiFePO_4$ [Gong and Yang 2011]. Therefore, for application purposes, the conductivity needs to be improved either by carbon coating or by synthesizing nanoparticles [Nitta et al. 2015]. The addition of a conductive phase is generally needed for improved performance. Other phosphates used for cathodes in LIBs include $LiMnPO_4$, $LiNiPO_4$, and $LiCoPO_4$. $LiMnPO_4$ shows a low capacity and poor cycling performance, whereas $LiNiPO_4$ and $LiCoPO_4$ have high operating voltage, which does not lie in the stability window of commercialized electrolytes, making these not suitable for practical applications. Table 3.1 summarizes various information and properties of few important cathode materials that have been used in LIBs [Manthiram 2011, Nitta et al. 2015, Manthiram 2004, Gong and Yang 2011, Pasquali 2004, Yoshio 2009, Julien et al. 2014, Islam and Fisher 2014].

TABLE 3.1

Properties Including Crystal Structure, Cell Voltage, Practical Capacity, Thermal Stability, the Dimensionality of Li-ion Diffusion, and Level of Development of Different Cathode Materials for LIBs

System	Structure	Cell voltage (V)	Practical capacity ($mAhg^{-1}$)	Thermal stability	Li-ion diffusion	Level of development
$LiCoO_2$	Layered$R\bar{3}m$	3.9	140	Fair	2D	Commercialized
$LiNiO_2$	Layered$R\bar{3}m$	4.0	150	Poor	2D	Research
$LiMnO_2$	Layered$R\bar{3}m$	3.3	140	Poor	2D	Research
NMC	Layered$R\bar{3}m$	3.7 (3.5 − 4.2)	150	Good	2D	Commercialized
$LiMn_2O_4$	Spinel $Fd\bar{3}m$	4.0	120	Good	3D	Commercialized
$LiFePO_4$	Olivine $Pnmb$	3.5	160	Excellent	1D [010] direction	Commercialized

Source: Manthiram 2011, Nitta et al. 2015, Manthiram 2004, Gong and Yang 2011, Pasquali 2004, Yoshio 2009, Julien et al. 2014, Islam and Fisher 2014.

3.4 CRYSTAL STRUCTURE AND ELECTRONIC PROPERTIES OF THREE CLASSES OF LITHIUM INTERCALATION-BASED CATHODE MATERIALS

The representative crystal structures of the three classes of lithium intercalation-based cathode materials are shown in Figure 3.1 [Nitta et al. 2015, Goodenough and Kim 2010, Islam and Fisher 2014, Tang et al. 2015]. $LiCoO_2$ is isostructural to layered α-$NaFeO_2$ with alternate Li^+ and $[CoO_2]^-$ layers. The Li^+ and Co^{3+} are octahedrally coordinated in a cubic close-packed (ccp) O^{2-} lattice resulting in a rhombohedral structure. The $LiMn_2O_4$ is isostructural to the cubic spinel AB_2O_4 type structure. In $LiMn_2O_4$, Li^+ ions are tetrahedrally coordinated, whereas Mn ions are octahedrally coordinated to oxygens. The Mn ions form a three-dimensional framework of edge-sharing MnO_6 octahedra, and Li tetrahedra share common faces with four neighboring empty octahedral sites [Nitta et al. 2015, Whittingham 2004, Islam and Fisher 2014, Tang et al. 2015, Thackeray 1995]. The olivine-type structure of $LiFePO_4$ consists of a distorted hexagonal close-packed (hcp) oxygen framework; Li and Fe ions are located in the half of octahedral sites and P ions in one-eighth of the tetrahedral sites. Li^+ ions form one-dimensional tunnels in the host structure along the [010] direction that runs parallel to the planes of corner-sharing FeO_6 octahedra [Nitta et al. 2015, Julien et al. 2014, Islam and Fisher 2014, Tang et al. 2015].

The oxygen evolution from lithium cobalt oxide triggers the thermal runaways in the case of oxides. Herein, the electronic properties of $LiCoO_2$, $LiMn_2O_4$, and $LiFePO_4$ have been assessed, and oxygen evolution can be correlated with the density of states of that material.

The oxygen loss behavior can be understood by considering the qualitative energy diagram of Li_xCoO_2 ($x = 1.0, 0.5, 0.0$) with the change in lithium concentration and is shown in Figure 3.2. In the case of $LiCoO_2$, the t_{2g} band is filled and e_g band is empty with the configuration of ($t_{2g}^{6}e_g^{0}$). On removal of lithium from $LiCoO_2$, the Co^{3+} is oxidized to Co^{4+} by removal of electrons from t_{2g} band. The energy

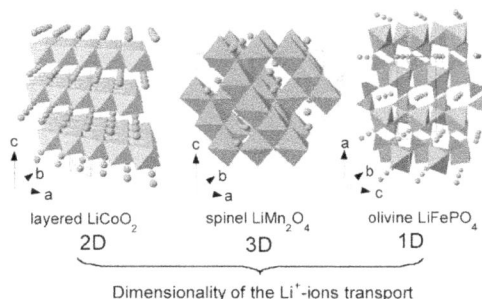

$\overset{c}{\underset{a}{\uparrow}} \overset{\nwarrow b}{}$ $\overset{c}{\underset{a}{\uparrow}} \overset{\nwarrow b}{}$ $\overset{a}{\underset{c}{\uparrow}} \overset{\nwarrow b}{}$

layered $LiCoO_2$ spinel $LiMn_2O_4$ olivine $LiFePO_4$
2D   3D   1D

Dimensionality of the Li^+-ions transport

FIGURE 3.1 Crystal structure of three commercialized lithium intercalation-based cathode materials: $LiCoO_2$ (layered), $LiMn_2O_4$ (spinel), and $LiFePO_4$ (olivine). Li-ions are shown as green spheres, CoO_6 octahedra in blue, MnO_6 octahedra in magenta, Fe polyhedra in brown, and PO_4 tetrahedra in purple. Black lines define one-unit cell in each structure [Reproduced with permission from Julien et al. 2014].

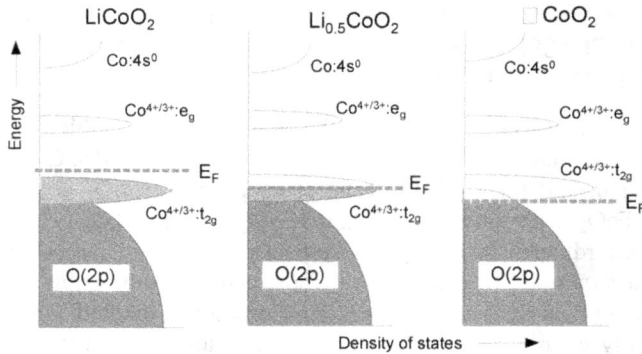

FIGURE 3.2 The qualitative energy diagrams of Li_xCoO_2 (x = 1.0, 0.5, 0.0) with the change of lithium concentration [Reproduced with permission from Julien et al. 2014].

diagram clearly shows that t_{2g} band overlaps with the top of O-2p band. The extraction of lithium beyond 0.5 might result in the creation of a hole in O-2p by removing electrons. The significant removal of electron density from O-2p band results in the oxidation of O^{2-}, which leads to the evolution of oxygen from lattice [Manthiram 2004, Julien et al. 2014, Chebiam et al. 2001].

Unlike $LiCoO_2$, spinel $LiMn_2O_4$ does not lose oxygen from lattice at deep lithium extraction. High spin Mn^{3+} ($t_{2g}^3 e_g^1$) configuration involves the removal of electrons from e_g band, which is located well above the top of O-2p band (Figure 3.3). Therefore, $LiMn_2O_4$ cathode is superior compared to $LiCoO_2$ based

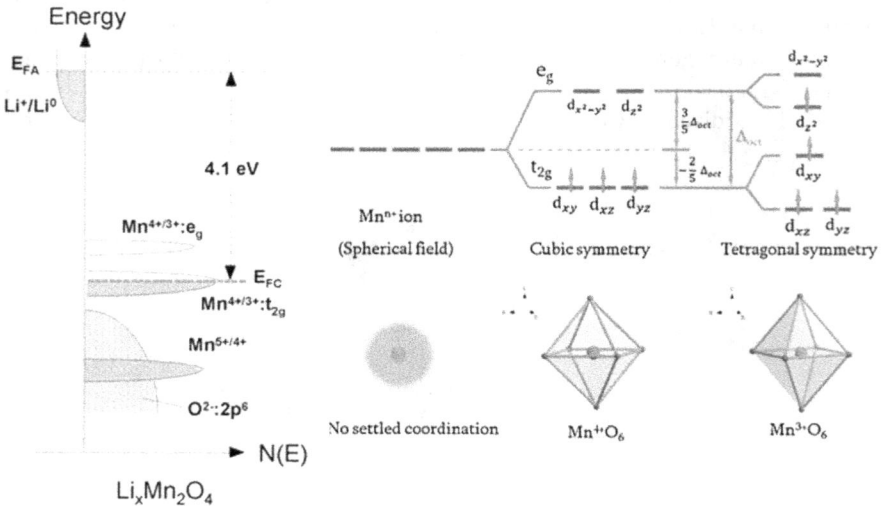

FIGURE 3.3 The qualitative schematic energy diagrams of $LiMn_2O_4$ along with the illustration of the Jahn–Teller distortion in manganese oxides associated with Mn^{3+}: $3d^4$ ($t_{2g}^3 e_g^1$) ion with tetragonal symmetry [Reproduced with permission from Julien et al. 2014, Liu et al. 2016].

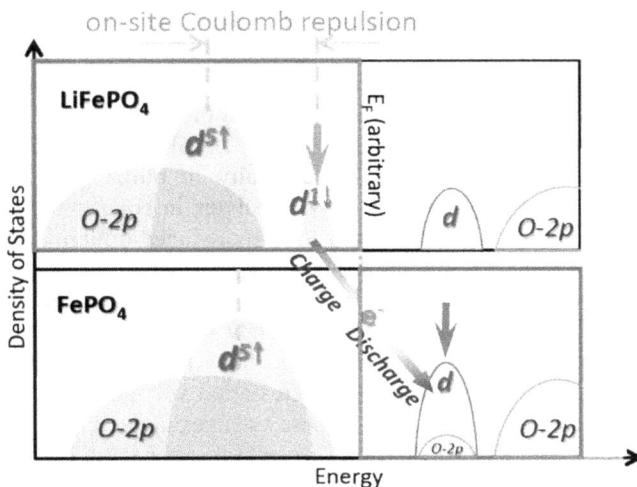

FIGURE 3.4 The schematic density of states in LiFePO$_4$and FePO$_4$ during the charge/discharge process. Electrons are transferred from the highest occupied d states of LiFePO$_4$ to the lowest unoccupied d states of FePO$_4$ [Reproduced with permission from Liu et al 2015].

on the resistance to oxygen loss from lattice [Julien et al. 2014, Chebiam et al. 2001], but it suffers from manganese dissolution and the Jahn–Teller distortion [Manthiram 2004]. An illustration of the Jahn–Teller distortion is shown in Figure 3.3, which is associated with the single electron in e$_g$ orbital of a high spin Mn^{3+}: 3d^4 (t$_{2g}$^{3}e$_g$^{1}) ion [Liu et al. 2016]. The stability of oxygen in LiFePO$_4$ during the charge/discharge process has been analyzed with a schematic density of states and is shown in Figure 3.4 [Liu et al 2015]. During charging, electrons are transferred from the highest occupied d states of LiFePO$_4$ to the lowest unoccupied d states of FePO$_4$. Due to the Coulomb interactions, the Fe-3d electrons with minority spins in LiFePO$_4$ are pushed up in energy and sit close to the Fermi level. Therefore, both the charging and discharging processes involve almost purely the Fe-3d states in the LiFePO$_4$ system leading to the safe performance of the electrodes during electrochemical operations [Liu et al 2015, Hafiz et al. 2017].

3.5 POLYOXYANION CATHODE MATERIALS

Polyanions are a relatively safer class of material compared to layered and spinels. Among polyanion cathode materials, LiFePO$_4$ is most extensively studied and considered to be one of the most promising cathode materials for hybrid electric vehicles [Goodenough 2007]. It has good structural and thermal stability and is made of nontoxic elements that are abundant in nature [Li et al. 2002]. However, many investigations are going on newer materials that show better performance and are environmentally benign. Polyoxyanion compounds (Li$_2$MSiO$_4$, M = Fe, Mn, Co, and Ni) are a new class of materials with high operating potential for second redox couples. The possibility of extraction of two lithium-ions from the Li_2MSiO_4 system results in higher capacity (above 330 $mAhg^{-1}$), leading to increased energy density [Manthiram 2011, Islam et al. 2011]. It can be observed that extraction of second lithium-ion will be at more than 4.5 V in all the four cases of Li_2MSiO_4 (M = Mn, Fe, Co, and Ni) [Arroyo-de Dompablo et al. 2006], which is the stability limit for $LiPF_6$-based electrolyte. The major difficulty in utilizing the high-voltage cathodes is the instability of the organic electrolytes in contact with the cathode surface at high operating voltages of >4.5 V. The predicted lithium de-intercalation voltages in the case of Li_2CoSiO_4 are 4.5 and 5.0 V for the first and second lithium, respectively, whereas 4.67 and 5.12 V for the first and second lithium, respectively, in the case of Li_2NiSiO_4, which are appropriately high in both the cases making these materials unsuitable for practical applications [Arroyo-de Dompablo et al. 2006]. In the case of Li_2MnSiO_4, the de-intercalation voltages for first and second lithium are 4.2 and 4.4 V, respectively. These voltages lie well within the stability window of $LiPF_6$-based electrolyte, resulting in higher capacity. But Li_2MnSiO_4 is predicted to be a poor Li-ion conductor at room temperature because of their high activation energies [Dominko et al. 2006, Dominko 2008] (at least 1 eV), which indicates low rate capability. Moreover, the rapid capacity fading can be seen for Li_2MnSiO_4-based cathodes [Duncan et al. 2011] because of the instability associated with the de-lithiated phase, making this material challenging for practical applications [Dominko 2008].

The lithium de-intercalation voltage in Li_2FeSiO_4 for first and second lithium occurs around ~3.1 and ~4.8 V, respectively. The value for first Li-ion de-intercalation is lowest in Li_2FeSiO_4 among all four silicates because it moved from a d^6 (Fe^{+2}) configuration to a closed-shell one d^5 (Fe^{+3}), which requires small ionization energy as compared to the other 3d M^{+2} to M^{+3} oxidations [Arroyo-de Dompablo et al. 2006]. Hence, the voltage step of 1.7 V between two voltage plateaus can also be predicted in Li_2FeSiO_4, which gets closer (~0.5 V) in other silicates (Li_2MSiO_4 with M = Mn, Co, and Ni) at 0 K. Li_2FeSiO_4 can be a promising material among all four silicates because silicon and iron are most abundant and low-cost elements, which results into sustainable, low-cost, and safe cathode material [Nitta et al. 2015, Islam et al. 2011].

Research activities have focused on the development of next-generation cathode and anode materials with enhanced capacity or cyclic performance. It is unlikely that significant advancement can be made soon because the specific capacity (in $mAhg^{-1}$) of the cathode material is often low (about half that of anode material). Theoretical calculations based on a simple relationship between the total capacity of the Li-ion battery material as a function of anode specific capacity (by using (3.1)) for the existing cathode materials are shown in Figure 3.5 [Sen et al. 2013].

$$\text{Total Capacity } C_{cell} = \frac{C_C \times C_A}{C_C + C_A} \qquad (3.1)$$

It can be seen from Figure 3.5 that for a fixed cathode capacity, the total specific capacity does not increase linearly with the linear increase in anode capacity. The total capacity increases quickly with an initial increase in anode capacity and then a

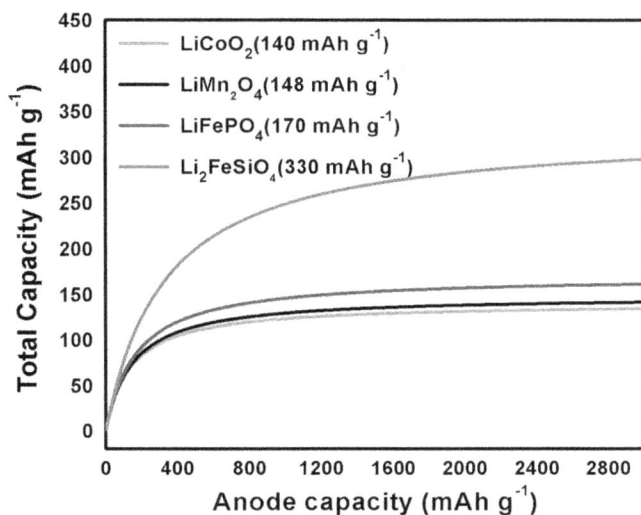

FIGURE 3.5 Variation of total cell capacity versus anode capacity for existing cathode materials [Reproduced with permission from Sen et al. 2013].

plateau-like region appears. For the existing cathode materials, the optimum specific capacity of the anode part is ranging from 1000 to 1200 mAhg^{-1}. The capacity contribution from the anode part is negligible after 1200 mAhg^{-1}. It also indicates that the total capacity increases substantially by just doubling cathode capacity. A good combination of cathode and anode is needed to demonstrate good cell performance for high-energy applications. Herein, Table 3.2 summarizes energy density for a possible combination of anode and cathode for some of the chosen materials. In the next section, several properties of Li_2FeSiO_4 material have been discussed including crystal structure and lithium diffusion pathways along with various advantages and disadvantages.

3.5.1 Polymorphism and Crystal Structure of Lithium Iron Silicate (Li_2FeSiO_4) Cathode Materials

Silicates also show polymorphism, which can be classified into low-temperature and high-temperature forms [Islam et al. 2011, Billaud et al. 2017]. All cations (Li^+, Fe^{2+}, and Si^{4+}) are tetrahedrally coordinated with oxygen in reported polymorphs [Zhang et al. 2012, Saracibar et al. 2012]. Li_2FeSiO_4 can be synthesized experimentally in three different space groups {$Pmn2_1$ [Abouimrane and Armand 2005], $Pmnb$ [Sirisopanaporn et al. 2010] (orthorhombic), and $P2_1/n$ [Zhang et al. 2009] (monoclinic)} by varying underlying experimental conditions such as calcination time and temperature [Islam et al. 2011, Sirisopanaporn et al. 2011a, 2011b]. The crystal structures of three very well-defined crystallographically pure Li_2FeSiO_4 are shown in Figure 3.6. The space group of crystallized Li_2FeSiO_4 is $Pmnb$ for LFS@ 900, $P2_1/n$ for LFS@700, and $Pmn2_1$ for LFS@200.

TABLE 3.2

Comparison of Full Cell Energy Density with Some of the Existing Cathode and Anode Materials

Cathode →	L iCoO$_2$[35] 3.9 V 140 mAh g^{-1}	LiMn$_2$O$_4$[36] 4.1 V 120 mAh g^{-1}	LiFePO$_4$[37] 3.44 V 160 mAh g^{-1}	Li$_2$FeSiO$_4$[38] 3.1 V 340 mAh g^{-1}
Anode↓	Full cell combination energy density			
Graphite [39] 0.1 V372 mAhg^{-1}	386 mWhg^{-1}	363 mWhg^{-1}	374 mWhg^{-1}	533 mWhg^{-1}
Graphene [40] 0.1 V740 mAhg^{-1}	447 mWhg^{-1}	413 mWhg^{-1}	439 mWhg^{-1}	699 mWhg^{-1}
TiO$_2$[41]1.5 V250 mAhg^{-1}	215 mWhg^{-1}	211 mWhg^{-1}	189 mWhg^{-1}	230 mWhg^{-1}
Fe$_2$O$_3$[42]0.8 V1000 mAhg^{-1}	380 mWhg^{-1}	353 mWhg^{-1}	364 mWhg^{-1}	583 mWhg^{-1}
MoS$_2$[43]0.5 V880 mAhg^{-1}	411 mWhg^{-1}	380 mWhg^{-1}	398 mWhg^{-1}	638 mWhg^{-1}
Si [44]0.4 V3000 mAhg^{-1}	468 mWhg^{-1}	427 mWhg^{-1}	462 mWhg^{-1}	824 mWhg^{-1}

Source: Reproduced with permission from Sen et al. 2013, Li et al. 2008, Jang et al. 1996, Kang and Cedar 2009, Rangappa et al. 2012, Kumar et al. 2009, Yoo et al. 2008, Liu et al. 2011, Wang et al. 2011a, Sen and Mitra 2013, Chan et al. 2008.

The differences in the local environments of Fe^{2+} and the interconnectivity of LiO$_4$ and SiO$_4$ tetrahedra in addition to their respective orientations along with a given crystallographic direction result in different crystal structures [Li et al. 2002, Saracibar et al. 2012, Eames et al. 2012]. The variations in the FeO$_4$ arrangements (orientation, size, and distortion) influence the equilibrium potential measured during the first oxidation of Fe^{2+} into Fe^{3+} in all polymorphs [Sirisopanaporn et al. 2011b, Eames et al. 2012]. The shorter Fe-O bonds result in the high splitting energy between bonding and antibonding states and thus lowering the Fe^{2+}/Fe^{3+} redox potential versus Li$^+$/Li^0. The Fe^{2+}/Fe^{3+} redox potential versus Li$^+$/Li^0 for each polymorph is investigated by potentiostatic intermittent titration technique (PITT) shown in Figure 3.6. The observed first oxidation potential is highest for *Pmn2$_1$*-based Li$_2$FeSiO$_4$ (3.13V versus Li$^+$/Li^0), whereas it is lowest for *Pmnb*-based Li$_2$FeSiO$_4$ (3.06 V versus Li$^+$/Li^0). The first oxidation potential value is 3.10 V versus Li$^+$/Li^0 for *P2$_1$/n*-based Li$_2$FeSiO$_4$. The difference in the first oxidation

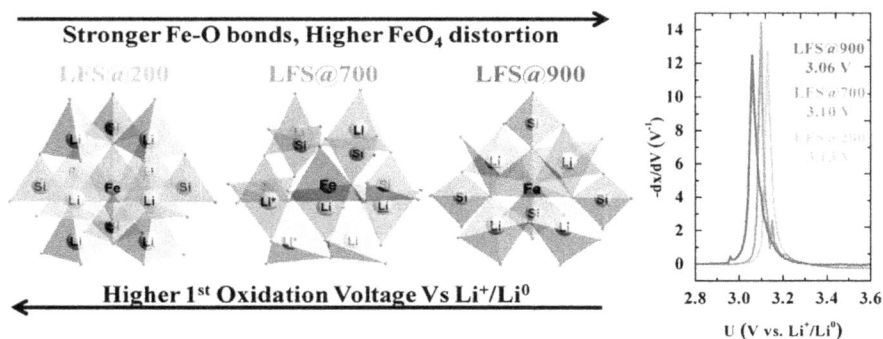

FIGURE 3.6 Local environments around FeO_4 tetrahedra (in green) in the three polymorphs of Li_2FeSiO_4 (LiO_4 in gray and SiO_4 in blue) and derivative plots obtained from PITT in the first oxidation of three polymorphs [Reproduced with permission from Sirisopanaporn et al. 2011b].

potential among the three polymorphs confirmed the influence of the crystallographic structures on redox potential [Sirisopanaporn et al. 2011b]. Moreover, the synthesis of pure phase is also challenging in silicate due to the very small differences in the formation energies of the three polymorphs [Gong et al. 2011]. Hence, the as-prepared samples usually occur as mixtures of two or even all three polymorphs [Gong et al. 2011] with detectable cation disorder that causes considerable difficulties in the structural refinement.

3.5.2 LI-ION DIFFUSION PATHWAYS

The transport pathways in Li_2FeSiO_4 have been characterized with computational studies depending on the polymorph structure, which may change during cycling. The diffusion direction for $Pmn2_1$ (orthorhombic) and $P2_1$ (monoclinic) based Li_2FeSiO_4 is briefly discussed in the subsequent sections. The lithium diffusion path for $Pmn2_1$-based Li_2FeSiO_4 has been investigated by Density-functional theory study [Armstrong et al. 2011] because probing of diffusion path is difficult by experiments [Islam et al. 2014]. Figure 3.7 shows two main lithium migration paths for cycled structures.

The first path shown in Figure 3.7(a) involves hopping between corner-sharing Li1 and Li2 sites with an overall trajectory along the c-axis direction. The second path [Figure 3.7(b)] involves hops between Li1 and Li2 sites but in the b direction with longer hop distances. The migration energy is 0.9 eV for the path in the c direction, which is much lower than the path along the b-axis (1.5 eV). This suggests that the favorable Li^+ migration path involves passage through intervening vacant octahedral sites that share faces with the tetrahedral Li1 and Li2 sites resulted in zigzag paths for Li transport [Armstrong et al. 2011, Islam et al. 2014].

The lithium diffusion pathway for $P2_1$-based Li_2FeSiO_4 (monoclinic) has also been studied. Figure 3.8 shows all possible pathways for Li-ion diffusion in the case

(a)   (b)

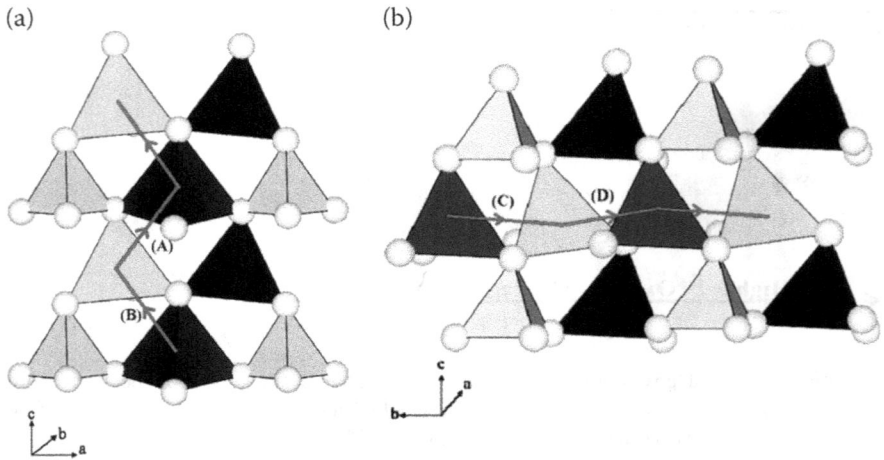

FIGURE 3.7 Pathways for lithium-ion migration between corner-sharing Li (1) and Li(2) sites in the cycled structure of Li_2FeSiO_4. (a) The first path involves hops A and B in the c direction. (b) The second path involves hops C and D in the b direction (SiO_4 tetrahedra, yellow; Li(1)O_4 tetrahedra, dark blue; Li(2)O_4 tetrahedra, light blue; FeO_4 tetrahedra, brown) [Reproduced with permission from Armstrong et al. 2011].

(a)   (b)

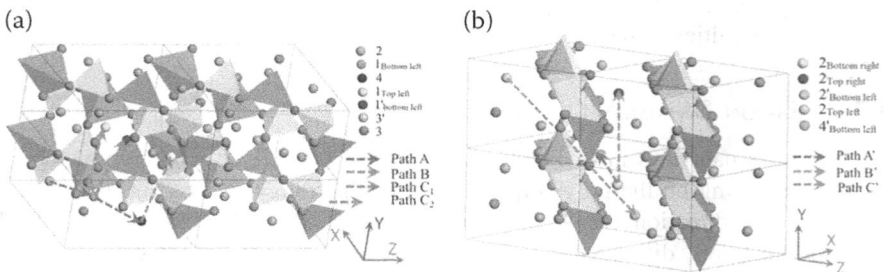

FIGURE 3.8 Li-vacancy migration pathways in the $P2_1$-based (a) Li_2FeSiO_4 and (b) $LiFeSiO_4$ structures. The red balls represent oxygen. Different migrations of Li ions are marked by number and position. The other purple balls represent Li ions, which are static during the migration. The yellow and blue tetrahedrons represent the SiO_4 and FeO_4, respectively [Reproduced with permission from Su et al. 2011].

of Li_2FeSiO_4 and its cycled structure. As shown in Figure 3.8(a), four migration pathways were identified in the case of Li_2FeSiO_4, which are as follows [Su et al. 2011]:

1. A: across channels along [101] direction.
2. B: a zigzag trajectory in the ac ($10\bar{1}$) crystal plane.
3. C1 and C2: through a small void in Fe-Si layer along [111] direction.

The activation barrier for the overall process in the case of the A pathway is 0.87 eV, and the value is 0.83 eV along the B pathway. The activation barrier for direct diffusion between adjacent sites is 1.99 and 2.51 eV for C1 and C2 pathways, respectively, which is substantially higher than the pathways A and B.

Figure 3.8(b) shows three migration pathways in the case of LiFeSiO$_4$:

1. A: across channels along [101] direction.
2. B: between adjacent M1 sites along the [010] direction (parallel to b-axis).
3. C: through the void in the Fe-Si layer.

The activation barriers are 0.87 and 0.79 eV for pathways A and B in the case of de-lithiated LiFeSiO$_4$. While the activation barrier for pathway C is 3.06 eV. From the above discussion, it can be concluded that migration energy is low along with A and B pathways in Li$_2$FeSiO$_4$ and LiFeSiO$_4$. Therefore, the diffusion in *P2$_1$* symmetry-based Li$_2$FeSiO$_4$ upon lithiation or de-lithiation is two-dimensional [Islam et al. 2014, Su et al. 2011]. The lithium diffusion pathways for *Pmnb* (orthorhombic) symmetry-based Li$_2$FeSiO$_4$ are not reported in current literature.

However, lower electronic/ionic conductivity is an inherent disadvantage of this material [Islam et al. 2011, Dominko 2008], which affects the electrochemical performance in various battery applications. Typical capacities obtained in initial reports were in the range of 120–140 mAh g^{-1} in the voltage range below 4 V versus Li [Abouimrane and Armand 2005]. However, claims have been made in the literature that the enhanced electrochemical activity can be achieved by few approaches: (i) providing proper channels for electrons and ions to interact inside the material by nanostructuring process [Huang et al. 2001, Gibot et al. 2008] and (ii) conductive carbon coating on electrode material [Gong et al. 2011, Huang et al. 2001, Gibot et al. 2008], which would benefit the electron transfer to the adjacent particles, thereby making an electronic bridge between them and hence reducing grain boundary impedance for mass and electron transfer. Above mentioned methods have been established for LiFePO$_4$ [Huang et al. 2001, Gibot et al. 2008], and similar approaches have been extended for Li$_2$FeSiO$_4$ to obtain enhanced electrochemical activity, which involve the addition of a carbon precursor (citrate anion, glucose, sucrose, ethylene glycol, etc.) to reduce the particle size and in situ carbon coating [Gong et al. 2011]. The presence of carbon during heat treatment suppresses the active particle growth, and particle agglomeration results in a reduction of particle size along with carbon coating. A few important reports in the literature on the electrochemical performance of Li$_2$FeSiO$_4$ and its dependence on the synthesis method, particle size, presence of an impurity, carbon percentage, and potential window of used electrolyte are tabulated in Table 3.3.

It can be seen from Table 3.3 that the obtained capacity mainly depends on the morphology, particle size, and synthesis route.

Synthesizing material by solution method results in nanomaterial that helps in achieving more capacity because Li diffusion path length reduces substantially in the case of nanomaterials [Arico et al. 2005, Bruce et al. 2008]. As demonstrated by Dinesh Rangappa and Itaru Honma et al., by synthesizing ultrathin sheets of Li$_2$FeSiO$_4$, successful extraction of two lithium at 45 °C up to 20 cycles is possible

TABLE 3.3

The Summary of Electrochemical Performance of Li_2FeSiO_4 Along with Synthesis Method, Particle Size, Presence of an Impurity, Carbon Percentage, and Potential Window of Electrolyte

Synthesis method	Space group	Particle size (nm)	Impurity	AM:C: PVDF& Potential window	Electrolyte	Discharge capacity (mAhg^{-1})
Solid-state reaction [Abouimrane and Armand 2005]	$Pmn2_1$	150	Li_2SiO_3, FeO_x	85:15:52.0-3.7 Vat 60 °C	1 M LiTFSI in EC:DEC (2:1)	140_C/16(10 cycles)
Solid-state reaction [Nytén et al. 2006]	$Pmn2_1$	150–200	$Fe_{1-x}OLiFeSi_2O_6$	80:10:102.0-3.7 V at 60 °C	1MLiTFSI in EC:PC (1:1)	125_C/25
Pechinisol–gel process [Dominko et al. 2006]	$Pmn2_1$	70	$Fe_3O_4Li_2SiO_3$	85:8:72.0-4.5V at RT&60 °C	1M LiPF$_6$ in EC:DMC (1:1)	55% at RT 85% at 60 °C_C/30
Hydrothermal synthesis [Dominko et al. 2008]	$Pmn2_1$	200–300	Fe (III) impurities	90:0:102.0-3.7 at 60 °C	0.8M LiBOBin EC:DEC (1:1)	60 %_C/20
Sol–gel method [Li et al. 2009]	$Pmn2_1$	80	Fe_3O_4	80:10:101.5-4.8 V	1M LiPF$_6$ in EC:DMC:EMC (1:1:1)	150_C/16100_2C
Microwave solvothermal [Muraliganth et al. 2010]	$P2_1$	150		75:20:51.5-4.8 V	1 M LiPF$_6$ in DC:EC (1:1)	148_RT204_55 °C
One-step solution polymerization [Lv et al. 2011]	$P2_1/n$	40–50	1.5-4.8 V at 30°C	85:5:101.5-4.8 V at 30°C	1M LiPF$_6$ in EC:DMC (1:1)	200_10 mAg^{-1}
Sol–gel method [Wu et al. 2012]	$P2_1/n$	30	1.5-4.8 V	80:15:151.5-4.8 V	1M LiPF$_6$ in EC:DMC (1:1)	230_0.1 C185_1 C
Supercritical fluid synthesis [Rangappa et al. 2012]	$Pmn2_1$	1–1.3 thickness Heights ~ 3 nm	Nanosheets	85:10:51.5-4.8 V	1 M LiClO$_4$ in EC:DEC (1:1)	160 at 19 °C, 340 at 45 ± 5 °C _C/50
Hydrothermal method [Yang et al. 2014]	$P2_1/n$	10–20	$Fe_3O_4Li_2SiO_3$	75:10:151.5-4.8 V	1 M LiPF$_6$ in EC:DMC (1:1)	211_0.1 C 189.8_0.5C
Sol–gel method [Tan et al. 2015]	$P2_1/n$	20–30, secondary particles 0.5–1 μm		1.5-4.5 V	PEO-Al$_2$O$_3$-LiTFSIElectrolyte (75:5:20)	258_1C (1.55 Li storage) at 100 °C

[Rangappa et al. 2012]. The obtained capacity of ~340 mAhg^{-1} by full lithium de-insertion has been assigned to electrochemically active Fe^{2+}/Fe^{3+} and Fe^{3+}/Fe^{4+} redox couples. Initially, the high discharge capacity, i.e. the capacity contribution beyond 166 mAhg^{-1} has been assigned to the electrochemically active Fe^{3+}/Fe^{4+} redox couple [Muraliganth et al. 2010, Wu et al. 2012, Rangappa et al. 2012]. The electrochemical activity of Fe^{3+}/Fe^{4+} redox couple was further confirmed by Lv et al. on analyzing ex situ Mossbauer spectroscopy and in situ X-ray absorption (XAS) experiments [Lv et al. 2014]. Later, Masese et al. have demonstrated that the formation of ligand holes in O-2p band is responsible for the extra capacity achieved during the second oxidation step for Li_2FeSiO_4, which compensates the charge after first lithium extraction [Masese et al. 2015]. In short, broadly two different mechanisms (i.e. electrochemically active Fe^{3+}/Fe^{4+} redox couple and O-2p hole formation) have been proposed in the literature for second lithium extraction, i.e. for obtained capacity of more than 166 mAhg^{-1}.

Anionic doping can be another strategy to enhance the electrochemical performance of a material by changing its electronic properties. For example, nitrogen (N), sulfur (S), fluorine (F), and chlorine (Cl) doping have been reported as effective dopants to improve the electrochemical performances of various electrode materials for LIBs [Armand and Arroyo-de Dompablo 2011, Zhu et al. 2015, Armand et al. 2011, Wang et al. 2011b]. The anionic (N, S, and F) doping has been very well studied theoretically in the case of Li_2FeSiO_4. It has been also observed that anionic doping can shift potential plateau voltage position, and thus, can help in tailoring material properties [Armand and Arroyo-de Dompablo 2011 2011, Zhu et al. 2015, Armand et al. 2011, Wang et al. 2011]. The substitution of N for O results in a decrease in the lithium de-insertion voltage, which results in improved electrochemical performance because the high operating voltage is associated with the second redox couple, i.e. Fe^{3+}/Fe^{4+} ~ 4.8 V also limits the extraction of second Li from the host matrix.

3.6 SYNTHESIS OF CATHODE MATERIAL

The first oil shock of 1976 had a massive impact in the field of inorganic chemistry synthesis with scientists preferring low-temperature approaches rather than ceramic-based methods [Schleich et al. 1994, Schollhorn 1988, Rouxel et al. 1996, Gopalakrishnan and J. Chimie Douce 1995] to reduce energy demands. Synthesis of the cathode material is critical to optimize the electrochemical behavior for commercial devices. The synthesized material needs to be well-characterized before using in devices. A well-characterized material can be tuned to use in battery applications. For example, its particle size and morphology will be optimized for maximum reactivity and minimum corrosivity, and side reactions, and it may be doped or coated to enhance the conductivity. Many soft chemistry techniques such as hydrothermal, ion exchange, and sol–gel can be used to synthesize desired material. Several critical synthesis parameters will be discussed in this section. The formation temperature of the cathode material determines the defect structure. The strict ordering of cations is desired in all the layered structures; otherwise, the diffusion is limited.

Several layered lithium metal dioxides are formed at a high-temperature reaction of carbonates in the air. The synthesis of $LiCoO_2$ requires a calcination temperature of 900 °C [Mizushima et al. 1980]. No measurable cobalt in the lithium layer is observed when $LiCoO_2$ is formed at optimum temperature. But at a low temperature around 400 °C results in the spinel phase, which means some Co is in the Li layer [Gummow et al. 1992]. In the first look, the X-ray diffraction patterns of layered and spinel look almost identical that can be distinguished by the c/a ratio, which is around 4.9 for spinel, whereas it is closer to 5.00 for rhombohedral layer structure. A second clue for layered structure is the existence of doublet in X-ray powder pattern at around 66 ° (2θ) associated with 108 and 110 reflections, whereas spinel has single reflection (440) in this region. In low-temperature synthesis, the partial occupancy of lithium sites by transition metal ions is found. The $LiNiO_2$ was similarly synthesized, and the X-ray diffraction suggests significant lattice disorder. $LiCoO_2$ formed hydrothermally at 230 °C from cobalt nitrate, lithium hydroxide, and hydrogen peroxide, which had rather poor cycling characteristics. As noted, neither NiO_2 nor CoO_2 has been synthesized directly because of their thermodynamic instability. The NiO_2 is inherently unsafe; therefore, to reduce the possibility of explosive interactions with the electrolyte, the additional element can be used to partially replace the nickel. The addition of trivalent ions such as boron or aluminum prevents the complete removal of all the lithium, thus limiting the overall average oxidation of the nickel. Alternatively, part of the nickel may be replaced by, for example, cobalt as in $LiNi_{1-y}Co_yO_2$, which forms a solid solution. A combination of these two approaches has been extensively studied.

For many years, the only way of synthesizing cathode materials has been the solid-state reaction. In this process, the core of the solid may not be reached by the viscous melt, especially if the solid reactant has large particles and limited porosity, which leads to incomplete reactions. Since 1990s, alternative methods of synthesis have been proposed to form a better mixture of the precursors and to reduce both time and temperature of reactions. Sol–gel synthesis is the most popular method. In sol–gel, a sol that converts into a gel is typically converted into desired ceramic by annealing. The sol–gel process allows a high degree of control over the physico-chemical characteristics of the material obtained with high purity and homogeneity. This approach allows one to obtain small particle size materials and, hence, faster reactions through faster diffusion of intercalating ions. Control of the pH and concentration of the chelating agent is required for optimized characteristics.

In the solvothermal and hydrothermal synthesis of cathode materials, the reactants are dissolved in water or in another solvent and then heated above the boiling temperature of the solvent for the desired length of time. The heating may be performed in either a conventional oven or a microwave oven. This technique has been used commercially for several decades to synthesize many materials including zeolites. It is only in the last two decades that much effort has been focused on transition metal compounds for battery applications. Hydrothermal and solvothermal approaches are appealing as they could produce phase-pure materials with unique morphologies. Materials can be obtained at <300 °C within a relatively short reaction time of 5–15 minutes. This approach not only decreases the manufacturing time but also allows the creation of unique nano-morphologies.

In the ionic liquid–assisted method, the ionic liquids are organic salts that are present in a liquid state at room temperature. Such ionic liquids have played the pivotal role of synthetic media in organic chemistry [Hulvey et al. 2009]. However, inorganic chemists have largely neglected the advantages of this reaction medium even though they have (i) very low vapor pressures, making them suitable for use in open reaction vessels and (ii) a decent solvating power making them suitable for a wide variety of precursors [Hulvey et al. 2009, Cooper et al. 2004]. Ionic liquids like organic solvents are easily extractable, separable, purifiable, and are recoverable and reusable. Due to the above-mentioned attributes, ionic liquids, namely, 1-ethyl-3-methylimidazolium bis (trifluoromethane-sulfonyl imide) (EMI-TFSI), have been prolifically used in preparing several different polyanionic electrode materials such as $LiFePO_4$, $LiMnPO_4$, $LiCoPO_4$, Na_2FePO_4F, and $LiFePO_4F$ [Tarascon et al. 2010]. It is worthwhile to note that like any other solvothermal method, this technique provides better control of particle size and morphology under atmospheric pressure and even at temperatures below 200–300 °C. The usage of ionothermal synthesis in inorganic synthesis is relatively nascent: substantial improvements can be expected in the future in terms of designing new low-cost ionic liquids for customized reaction conditions. Various options to replace imidazolium cations and highly expensive TFSI anions by quaternary ammonium cations and chloride anions, respectively, are presently being explored and also not compromising on overall stability. Ionic liquids can be a renewable alternative to other low-boiling solvents.

New low-cost processing initiatives should ideally give rise to effective synthetic approaches that can be performed at room temperature; biosynthesized materials have been conceptualized as electrode materials for LIBs.

In the biological synthesis route, materials are synthesized by biomineralization reactions performed by bacteria. $LiFePO_4$ has been synthesized in the presence of the bacteria *Bacillus pasteurii*. Enzyme urease is slowly released by these bacteria. It is then combined with stoichiometric amounts of LiH_2PO_4, $FeSO_4.H_2O$, and urea to produce the basic pH needed for the precipitation of $LiFePO_4$. The reactions are performed in aerobic conditions in sealed vessels at 60 °C; however, reproducibility of results is difficult due to challenges in maintaining the performance of the bacteria. These approaches are presently being applied to the synthesis of silicates and other phosphates and have shown encouraging results. There still exist major challenges in using this method; however, interdisciplinary collaboration with biologists will result in the production of large-scale reproducible results [Brent et al. 2013].

3.7 CATHODE MATERIALS AND BATTERY SAFETY

The safety of a battery is broadly divided into three categories: overcharging, safety at high temperature, and short-circuit. The charging of a battery beyond its working voltage is referred to as overcharging, which can be caused by the malfunctioning of the charger. Heat generation occurs with abnormal electrochemical reactions within the battery, while short-circuiting arises from manufacturing defects or battery abuse. These reactions may have different causes but are closely related to heat production and contribute to explosive heat reactions in the anode. Some related processes are the surface reaction between the electrolyte and the cathode, the pyrolysis reaction of the cathode (oxygen production), the oxidation reaction of the electrolyte, and the pyrolysis reaction of the electrolyte. Since the electrolyte has a high decomposition potential than the cathode, therefore, normal conditions do not lead to electrolyte decomposition. The overcharged state, when the potential of the cathode rises beyond the electrolyte decomposition potential, results in an oxidation reaction of the electrolyte accompanied by heat generation. Thermal decomposition of the cathode releases a great amount of heat and oxygen during the structural change in transition oxides used as cathode materials at high temperatures. Depending on the active material, the initial decomposition temperature is in the ascending order of $LiNiO_2 < LiCoO_2 < LiMn_2O_4$. The thermal stability of a charged cathode is used as a scale to measure battery safety, which is related to the structural stability of the cathode material. The cathode material exists in a stable state after synthesis and when discharged. However, it becomes thermodynamically unstable and enters a metastable state during the charging process. The heat is likely to be produced from a change in the structure of cathode active materials. Hence, the structural stability of cathode materials should be secured in the charged state. Thermal analysis of the charged product without electrolyte gives information about the temperature and amount of oxygen release. Dahn et al. have evaluated the heat generation of various cathodes in the presence of electrolytes by DSC (Differential scanning calorimetry) measurement. DSC is a very useful technique for comparing the behavior of electrodes under the conditions of thermal runaway, thus quantifying the relative safety of the electrodes. The total heat generation increases with an increase in the charge capacity for 4 V class cathode materials. The 3 V class $LiFePO_4$ shows low heat generation due to the stability of a strong covalent bond in PO_4^{3-}. In $LiNiO_2$, a sudden increase in heat generation can be seen at the $Li_{0.3}NiO_2$, due to the presence of the NiO_2 phase, which can be decreased by cobalt doping. On comparing, the thermal analysis for $Li_x[Ni_{1/3}Mn_{1/3}Co_{1/3}]O_2$ and Li_xCoO_2 shows that $Li_x[Ni_{1/3}Mn_{1/3}Co_{1/3}]O_2$ has a higher exothermic onset point and a smaller peak compared to Li_xCoO_2, although both have the same structure, $Li[Ni_{1/3}Mn_{1/3}Co_{1/3}]O_2$ is more stable than $LiCoO_2$. The more lithium is de-intercalated from $LiNiO_2$ due to its high capacity, making it more unstable than the $Li[Ni_{1/3}Mn_{1/3}Co_{1/3}]O_2$. Depending on the composition, layered oxide cathodes have been reported to have an enthalpy of reaction that typically ranges from over 1000 to 3300 J/g. The exothermic heat flow for $LiFePO_4$ is 210 J/g. The exothermic heat flow calculated from DSC is 330 J/g in the case of $Li_{2-x}FeSiO_4$ [Muraliganth et al. 2010]. The spinel $LiMn_2O_4$ is thermodynamically stable in both the charged and the discharged states with no heat production from structural change. The structure of $LiFePO_4$ is not changing by charging or heating; the active material remains unchanged up to 230 °C. Therefore, $LiFePO_4$ appears to be an extremely stable cathode material. The $Li_{2-x}MnSiO4$ has poor thermal stability, which is due to the structural instability of its de-lithiated phase. The thermal stability of Li_2FeSiO_4 is slightly worse than that of $LiFePO_4$, but it is much better than that of layered cathodes. Hence, $LiFePO_4$ and Li_2FeSiO_4 are attractive candidates for large-scale applications due to their good thermal stability and safety characteristics (Table 3.4).

TABLE 3.4
Some Available Patents on Cathodes

Inventor	Cathode	Patent No.	Year	Invention
John B. Goodenough et al.	$LiCoO_2$	EP 0017400A1	1980	$LiCoO_2$ having an α-$NaCrO_2$ structure where Li cations are extracted electrochemically by charging a cell represented as Li/electrolyte/$LiCoO_2$.
J. M. Tarascon	$Li_xMn_2O_4$ ($0 \leq x \leq 2$)	US 005,196,279	1993	About 0.8 lithium atoms per formula unit can be reversibly intercalated at an average potential of 3.7 volts while cycling between 4.5 and 2 V.
M. Thackeray et al.	$LiMn_2O_4$	US 005,316,877	1994	A cell of the type Li/1M $LiClO_4$ in $PC/LiMn_2O_4$ showed a cycling capacity of 115 mAhg^{-1}, which decreases with cycling. The capacity loss is attributed to tetragonal distortion due to the Jahn–Teller effect.
John B. Goodenough et al.	$LiFePO_4$	US 005910382 A	1999	Transition metal compounds having ordered olivine or the rhombohedral NASICON structure with at least one constituent $(PO_4)^{3-}$ are used as electrode material for LIBs.
Michel Armand et al.	$LiFePO_4$	US 6,514,640 B1	2003	The $LiFePO_4$ material of the present invention represents a cathode of good capacity and contains inexpensive, environmentally benign elements. A voltage of almost 3.5 V versus lithium for a capacity of 0.61 Li/formula unit at a current density of 0.05 mA.cm^{-2}.
S. Gopukumar et al.	$LiMg_xCu_yCo_{x-1-y}O_2$ ($0 \leq x,y \leq 0.2$)	US2014/ 0087257 A1	2014	The cathode is a dual doped lithium cobalt oxide of general formula $LiMg_xCu_yCo_{1-x-y}O_2$ ($0 \leq x,y \leq 0.2$) is synthesized. A LIB of 2016 coin cell configuration delivers a discharge capacity of 200–240 mAhg^{-1} at 0.2 C rate with cycling stability of 90–97% after 100 cycles and discharge capacity of 70–90 mAh g^{-1} at 1 C rate with cycling stability of 95–98% after 200 cycles.
Xiao et al.	Li_2FeSiO_4	US 2020/ 0220172 A1	2020	The battery has an areal capacity of at least 3.7 mAh/cm^2 and can be charged to 90% of its full capacity within 10 minutes of charging. The battery is capable of a charge rate greater than or equal to about 4 C at 25°C.

3.8 SUMMARY

A brief overview of the major classes of cathode materials is presented in this chapter along with the essential characteristics of efficient cathode material. We have focused in particular on a detailed description of crystal structure and electronic properties of cathode materials, and how a change of lithium concentration affecting the thermodynamic and structural properties of the compound. The polymorphism in the case of lithium iron silicate is also discussed. The pathways for lithium-ion diffusion are also discussed whether it is one-, two- or three-dimensional diffusion. We have also discussed different synthesis techniques in brief. At last, the relative thermal stability of different cathode materials has been compared.

REFERENCES

Abouimrane, A., Armand, M. 2005. *Electrochemical Performance of Li_2FeSiO_4 as a New Li-Battery Cathode Material*, 7, 156–160.

Arico, A.S., Bruce, P., Scrosati, B., Tarascon, J.-M., Schalkwijk, W. 2005. Nanostructured Materials for Advanced Energy Conversion and Storage Devices. *Nat. Mater.*, 43 (6), 366–377.

Armand, M., Arroyo-de Dompablo, M.E. 2011. Benefits of N for O Substitution in Polyoxoanionic Electrode Materials: A First Principles Investigation of the Electrochemical Properties of $Li_2FeSiO_{4-y}N_y$ (y = 0, 0.5, 1). *J. Mater. Chem.*, 21 (27), 10026–10034.

Armand, M., Tarascon, J.-M., Arroyo-de Dompablo, M.E. 2011. Comparative Computational Investigation of N and F Substituted Polyoxoanionic Compounds The Case of Li_2FeSiO_4 Electrode Material. *Electrochem. Commun.*, 13 (10), 1047–1050.

Armstrong, A.R., Kuganathan, N., Islam, M.S., Bruce, P.G. 2011. Structure and Lithium Transport Pathways in Li_2FeSiO_4 Cathodes for Lithium Batteries. *J. Am. Chem. Soc.*, 133 (33), 13031–13035.

Arroyo-de Dompablo, M.E., Armand, M., Tarascon, J.M., Amador, U. 2006. On-Demand Design of Polyoxianionic Cathode Materials Based on Electronegativity Correlations: An Exploration of the Li_2MSiO_4 System (M = Fe, Mn, Co, Ni). *Electrochem. Commun.*, 8 (8), 1292–1298.

Billaud, J., Eames, C., Tapia-Ruiz, N., Roberts, M.R., Naylor, A.J., Armstrong, A.R., Islam, M.S., Bruce, P.G. 2017. Evidence of Enhanced Ion Transport in Li-Rich Silicate Intercalation Materials. *Adv. Energy Mater.*, 7 (11), 1–9.

Brent, C. Melot and Tarascon, J.-M. 2013. Design and Preparation of Materials for Advanced Electrochemical Storage. *Acc. Chem. Res.*, 46 (5), 1226–1238.

Bruce, P.G., Scrosati, B., Tarascon, J.M. 2008. Nanomaterials for Rechargeable Lithium Batteries. *Angew. Chemie Int. Ed.*, 47 (16), 2930–2946.

Chan, C.K., Peng, H.L., Liu, G., McIlwrath, K., Zhang, X.F., Huggins, R.A., Cui, Y. 2008. High-performance lithium battery anodes using silicon nanowires. *Nat. Nanotechnol.*, 3, 31–35.

Chebiam, R., Kannan, A., Prado, F., Manthiram, A. 2001. Comparison of the Chemical Stability of the High Energy Density Cathodes of Lithium-Ion Batteries. *Electrochem. Commun.*, 3 (11), 624–627.

Cooper, E.R., Andrews, C.D., Wheatley, P.S., Webb, P.B., Wormald, P., Morris, R.E. 2004. Ionic Liquids and Eutectic Mixtures as Solvent and Template in Synthesis of Zeolite Analogues. *Nature*, 430, 1012–1016.

Dominko, R. 2008. Li_2MSiO_4 (M = Fe And/or Mn) Cathode Materials. *J. Power Sources*, 184 (2), 462–468.

Dominko, R., Bele, M., Gaberšček, M., Meden, A., Remškar, M., Jamnik, J. 2006. Structure and Electrochemical Performance of Li_2MnSiO_4 and Li_2FeSiO_4 as Potential Li-Battery cathode Materials. *Electrochem. Commun.*, 8 (2), 217–222.

Dominko, R., Conte, D.E., Hanzel, D., Gaberscek, M., Jamnik, J. 2008. Impact of Synthesis Conditions on the Structure and Performance of Li_2FeSiO_4. *J. Power Sources*, 178 (2), 842–847.

Duncan, H., Kondamreddy, A., Mercier, P.H.J., Le Page, Y., Abu-Lebdeh, Y., Couillard, M., Whitfield, P.S., Davidson, I.J. 2011. Novel Pn Polymorph for Li_2MnSiO_4 and Its Electrochemical Activity As a Cathode Material in Li-Ion Batteries. *Chem. Mater.*, 23 (24), 5446–5456.

Dunn, B., Dunn, B., Kamath, H., Tarascon, J. 2011. Electrical Energy Storage for the Gridfor the Grid: A Battery of Choices. *Sci. Mag.*, 334 (6058), 928–936.

Eames, C., Armstrong, A.R., Bruce, P.G., Islam, M.S. 2012. Insights into Changes in Voltage and Structure of Li_2FeSiO_4 Polymorphs for Lithium-Ion Batteries. *Chem. Mater.*, 24, 2155–2161.

Gibot, P., Casas-Cabanas, M., Laffont, L., Levasseur, S., Carlach, P., Hamelet, S., Tarascon, J.M., Masquelier, C. 2008. Room-Temperature Single-Phase Li insertion/extraction in Nanoscale Li_xFePO_4. *Nat. Mater.*, 7 (9), 741–747.

Gong, Z., Yang, Y. 2011. Recent Advances in the Research of Polyanion-Type Cathode Materials for Li-Ion Batteries. *Energy Environ. Sci.*, 4 (9), 3223.

Goodenough, J.B. 2007. Cathode Materials: A Personal Perspective. *J. Power Sources*, 174 (2), 996–1000.

Goodenough, J.B., Kim, Y. 2010. Challenges for Rechargeable Li Batteries. *Chem. Mater.*, 22 (3), 587–603.

Gopalakrishnan, J. Chimie Douce 1995. Approaches to the Synthesis of Metastable Oxide Materials. *Chem. Mater.*, 7, 1265–1275.

Hafiz, H., Suzuki, K., Barbiellini, B., Orikasa, Y., Callewaert, V., Kaprzyk, S., Itou, M., Yamamoto, K., Yamada, R., Uchimoto, Y., Sakurai, Y., Sakurai, H., Bansil, A. 2017. Visualizing Redox Orbitals and Their Potentials in Advanced Lithium-Ion Battery Materials Using High-Resolution X-Ray Compton Scattering. *Sci. Adv.*, 3 (8), e1700971.

Huang, H., Yin, S.-C., Nazar, L.F. 2001. Approaching Theoretical Capacity of $LiFePO_4$ at Room Temperature at High Rates. *Electrochem. Solid-State Lett.*, 4 (10), A170.

Hulvey, Z., Wragg, D.S., Lin, Z., Morris, R.E., Cheetham, A.K. 2009. Ionothermal synthesis of inorganic-organic hybrid materials containing per fluorinated aliphatic dicarboxylate ligands. *Dalton Trans.*, 7, 1131–1135.

Islam, M.S., Dominko, R., Masquelier, C., Sirisopanaporn, C., Armstrong, A.R., Bruce, P.G. 2011. Silicate Cathodes for Lithium Batteries: Alternatives to Phosphates? *J. Mater. Chem.*, 21 (27), 9811.

Islam, M.S., Fisher, C.A.J. 2014. Lithium and Sodium Battery Cathode Materials: Computational Insights into Voltage, Diffusion and Nanostructural Properties. *Chem. Soc. Rev.*, 43 (1), 185–204.

Jang, D.H., Shin, Y.J., Oh, S.M. 1996. Dissolution of Spinel Oxides and Capacity Losses in 4V $Li/Li_xMn_2O_4$ Cells. *J. Electrochem. Soc.*, 143, 2204–2211.

Julien, C.M., Mauger, A., Zaghib, K., Groult, H. 2014. Comparative Issues of Cathode Materials for Li-Ion Batteries. *Inorganics*, 2, 132–154.

K. Mizushima, P.C. Jones, P.J. Wiseman, J.B. Goodenough. 1980. $LixCoO_2$ (0<x<-1): A New Cathode Material for Batteries of High Energy Density. *Mater. Res. Bull.*, 15 (6), 783–789.

Kang, B., Ceder, G. 2009. Battery Materials for Ultrafast Charging and Discharging. *Nature*, 458, 190–193.

Kumar, T.P., Kumari, T.S.D., Stephan A.M. 2009. Carbonaceous anode materials for lithium-ion batteries-the road ahead. *J. Ind. Inst. Sci.*, 89, 393–424.

Li, G., Azuma, H., Tohda, M. 2002. Optimized $LiMn_yFe_{1-y}PO_4$ as the Cathode for Lithium Batteries. *J. Electrochem. Soc.*, 149 (6), A743.

Li, G., Yang, Z., Yang, W. 2008. Effect of $FePO_4$ coating on electrochemical and safety performance of $LiCoO_2$ as cathode for Li-ion batteries. *J. Power Sources*, 183, 741–748.

Li, L. Ming, Guo, H. Jun, Li, X. Hai, Wang, Z. Xing, Peng, W. Jie, Xiang, K. Xiong, Cao, X. 2009. Effects of Roasting Temperature and Modification on Properties of Li_2FeSiO_4/C Cathode. *J. Power Sources*, 189 (1), 45–50.

Liu, C., Neale, Z.G., Cao, G. 2016. Understanding Electrochemical Potentials of Cathode Materials in Rechargeable Batteries. *Mater. Today*, 19 (2), 109–123.

Liu, H., Bi, Z., Sun, X.-G., Unocic, R.R., Paranthaman, M.P., Dai, S., Brown, G.M. 2011. Mesoporous TiO_2-B Microspheres with Superior Rate Performance for Lithium Ion Batteries. *Adv. Mater.*, 23, 3450–3454.

Liu, X., Wang, Y.J., Barbiellini, B., Hafiz, H., Basak, S., Liu, J., Richardson, T., Shu, G., Chou, F., Weng, T.-C., Nordlund, D., Sokaras, D., Moritz, B., Devereaux, T.P., Qiao, R., Chuang, Y.-D., Bansil, A., Hussain, Z., Yang, W. 2015. Why $LiFePO_4$ Is a Safe Battery Electrode: Coulomb Repulsion Induced Electron-State Reshuffling upon Lithiation. *Phys. Chem. Chem. Phys.*, 17 (39), 26369–26377.

Lv, D., Bai, J., Zhang, P., Wu, S., Li, Y., Wen, W., Jiang, Z., Mi, J., Zhu, Z., Yang, Y. 2014. Understanding the High Capacity of Li_2FeSiO_4: In Situ XRD/XANES Study Combined with First-Principles Calculations. *Chem. Mater.* 25 (10), 2014–2020.

Lv, D., Wen, W., Huang, X., Bai, J., Mi, J., Wu, S., Yang, Y. 2011. A Novel Li_2FeSiO_4/C Composite: Synthesis, Characterization and High Storage Capacity. *J. Mater. Chem.*, 21 (26), 9506.

Manthiram, A. 2004. Materials Aspects: An Overview. In *Lithium Batteries - Science and Technology*, Nazri, G., Pistoia, G., Ed. Kluwer Academic Publishers.

Manthiram, A. 2011. Materials Challenges and Opportunities of Lithium Ion Batteries. *J. Phys. Chem. Lett.* 2 (3), 176–184.

Masese, T., Tassel, C., Orikasa, Y., Koyama, Y., Arai, H., Hayashi, N., Kim, J., Mori, T., Yamamoto, K., Kobayashi, Y., Kageyama, H., Ogumi, Z., Uchimoto, Y. 2015. Crystal Structural Changes and Charge Compensation Mechanism during Two Lithium Extraction/insertion between Li_2FeSiO_4 and $FeSiO_4$. *J. Phys. Chem. C*, 119 (19), 10206–10211.

Muraliganth, T., Stroukoff, K.R., Manthiram, A. 2010. Microwave-Solvothermal Synthesis of Nanostructured Li_2MSiO_4/C (M = Mn and Fe) Cathodes for Lithium-Ion Batteries. *Chem. Mater.*, 22 (20), 5754–5761.

Nazar, L.F. 2004. Anodes and Composite Anodes: An Overview. In *Lithium Batteries - Science and Technology*, Pistoia, G.-A.N.G., Ed. Kluwer Academic Publishers.

Nitta, N., Wu, F., Lee, J.T., Yushin, G. 2015. Li-Ion Battery Materials: Present and Future. *Mater. Today*, 18 (5), 252–264.

Nytén, A., Kamali, S., Häggström, L., Gustafsson, T., Thomas, J.O. 2006. The Lithium Extraction/insertion Mechanism in Li_2FeSiO_4. *J. Mater. Chem.*, 16 (23), 2266–2272.

Ohzuku, T., Makimura, Y. 2001. Layered Lithium Insertion Material of $LiCo_{1/3}Ni_{1/3}Mn_{1/3}O_2$ for Lithium-Ion Batteries. *Chem. Lett.*, 30 (7), 642–643.

Pasquali, M. 2004. Trends in Cathode Materials for Rechargeable Batteries. In *Lithium Batteries - Science and Technology*, Pistoia, G.-A.N.G., Ed. Kluwer Academic Publishers.

Rangappa, D., Murukanahally, K.D., Tomai, T., Unemoto, A., Honma, I. 2012. Ultrathin Nanosheets of Li_2MSiO_4 (M = Fe, Mn) as High-Capacity Li-Ion Battery Electrode. *Nano Lett.*, 12 (3), 1146–1151.

R.J. Gummow, M.M. Thackeray, W.I.F. David, S. Hull 1992. Structure and electrochemistry of lithium cobalt oxide synthesised at 400°C. *Mater. Res. Bull.*, 27 (3), 327–337.

Rouxel, J., Tournoux, M. 1996. Chimie Douce with Solid Precursors, Past and Present. *Solid State Ionics*, 84, 141–149.

Saracibar, A., Van der Ven, A., Arroyo-de Dompablo, M.E. 2012. Crystal Structure, Energetics, And Electrochemistry of Li_2FeSiO_4 Polymorphs from First Principles Calculations. *Chem. Mater.*, 24 (3), 495–503.

Schleich, D.M. 1994. Chimie Douce: Low Temperature Techniques for Synthesizing Useful Compounds. *Solid State Ionics*, 7071, 407–411.

Schollhorn, R. 1988. From Electronic/Ionic Conductors to Superconductors: Control of Materials Properties. *Angew. Chem. Int. Ed.*, 27, 1392–1400.

Sen, U.K., Mitra, S. 2013. High Rate and High Energy Density Lithium-Ion Battery Anode Containing 2D MoS_2 Nanowall and Cellulose Binder. *ACS Appl. Mater. Interfaces*, 5, 1240–1247.

Sen, U.K., Sarkar, S., Veluri, P.S., Singh, S. 2013. Nano Dimensionality: A Way Towards Better Li-Ion Storage. *Nanosci. Nanotechnol. Asia*, 3, 21.

Sirisopanaporn, C., Boulineau, A., Hanzel, D., Dominko, R., Budic, B., Armstrong, A.R., Bruce, P.G., Masquelier, C. 2010. Crystal Structure of a New Polymorph of Li_2FeSiO_4. *Inorg. Chem.*, 49 (16), 7446–7451.

Sirisopanaporn, C., Dominko, R., Masquelier, C., Armstrong, A.R., Mali, G., Bruce, P.G. 2011a. Polymorphism in $Li_2(Fe,Mn)SiO_4$: A Combined Diffraction and NMR Study. *J. Mater. Chem.*, 21 (44), 17823–17831.

Sirisopanaporn, C., Masquelier, C., Bruce, P.G., Armstrong, A.R., Dominko, R. 2011b. Dependence of Li_2FeSiO_4 Electrochemistry on Structure. *J. Am. Chem. Soc.*, 133 (5), 1263–1265.

Su, D., Ahn, H., Wang, G. 2011. Ab Initio Calculations on Li-Ion Migration in Li_2FeSiO_4 Cathode Material with a $P2_1$ Symmetry Structure. *Appl. Phys. Lett.*, 99 (14), 2–5.

Tan, R., Yang, J., Zheng, J., Wang, K., Lin, L., Ji, S., Liu, J., Pan, F. 2015. Fast Rechargeable All-Solid-State Lithium Ion Batteries with High Capacity Based on Nano-Sized Li_2FeSiO_4 Cathode by Tuning Temperature. *Nano Energy*, 16, 112–121.

Tang, Y., Zhang, Y., Li, W., Ma, B., Chen, X. 2015. Rational Material Design for Ultrafast Rechargeable Lithium-Ion Batteries. *Chem. Soc. Rev.*, 44 (17), 5926–5940.

Tarascon, J.-M., Recham, N., Armand, M., Chotard, J.-N., Barpanda, P., Walker, W., Dupont, L. 2010. Hunting for Better Li-Based Electrode Materials via Low Temperature Inorganic Synthesis. *Chem. Mater.*, 22, 724–739.

Thackeray, M.M. 1995. Lithiated Oxides for Lithium Ion Batteries. *J. Electrochem. Soc.*, 142 (8), 2558–2563.

Wang, Z., Luan, D., Madhavi, S., Li, C.M., Lou, X.W. 2011a. α-Fe_2O_3 Nanotubes with Superior Lithium Storage Capability. *Chem. Commun.*, 47, 8061–8063.

Wang, Z.-H., Yuan, L.-X., Wu, M., Sun, D., Huang, Y.-H. 2011b. Effects of Na^+ and Cl^- Co-Doping on Electrochemical Performance in $LiFePO_4$/C. *Electrochim. Acta*, 56, 8477–8483.

Whittingham, M.S. 2004. Lithium Batteries and Cathode Materials. *Chem. Rev.*, 104 (10), 4271–4301.

Wu, X., Jiang, X., Huo, Q., Zhang, Y. 2012. Facile Synthesis of Li_2FeSiO_4/C Composites with Triblock Copolymer P123 and Their Application as Cathode Materials for Lithium Ion Batteries. *Electrochim. Acta*, 80, 50–55.

Xia, H., Luo, Z., Xie, J. 2012. Nanostructured $LiMn_2O_4$ and Their Composites as High-Performance Cathodes for Lithium-Ion Batteries. *Prog. Nat. Sci. Mater. Int.*, 22 (6), 572–584.

Xia, Y. 1997. Capacity Fading on Cycling of 4 V Li/$LiMn_2O_4$ Cells. *J. Electrochem. Soc.*, 144 (8), 2593.

Yamada, A., Chung, S.C., Hinokuma, K. 2001. Optimized $LiFePO_4$ for Lithium Battery Cathodes. *J. Electrochem. Soc.*, 148 (3), A224.

Yang, J., Kang, X., Hu, L., Gong, X., Mu, S. 2014. Nanocrystalline-Li_2FeSiO_4 Synthesized by Carbon Frameworks as an Advanced Cathode Material for Li-Ion Batteries. *J. Mater. Chem. A*, 2 (19), 6870–6878.

Yoo, E., Kim, J., Hosono, E., Zhou, H., Kudo, T., Honma, I. 2008. Large Reversible Li Storage of Grapheme Nanosheet Families for use in Rechargeable Lithium Ion Batteries. *Nano Lett.*, 8, 2277–2282.

Yoshio, M.N.H. 2009. A Review of Positive Electrode Materials for Lithium-Ion Batteries. In *Lithium-Ion Batteries Science and Technologies*, Yoshio, M., Brodd, R.J., Kozawa, A., Ed. Springer.

Yuan, L.-X., Wang, Z.-H., Zhang, W.-X., Hu, X.-L., Chen, J.-T., Huang, Y.-H., Goodenough, J.B. 2011. Development and Challenges of $LiFePO_4$ Cathode Material for Lithium-Ion Batteries. *Energy Environ. Sci.*, 4 (2), 269–284.

Zhang, P., Hu, C.H., Wu, S.Q., Zhu, Z.Z., Yang, Y. 2012. Structural Properties and Energetics of Li_2FeSiO_4 Polymorphs and Their Delithiated Products from First-Principles. *Phys. Chem. Chem. Phys.*, 14 (20), 7346–7351.

Zhang, S., Deng, C., Yang, S. 2009. Preparation of Nano-Li_2FeSiO_4 as Cathode Material for Lithium-Ion Batteries. *Electrochem. Solid-State Lett.*, 12 (7), A136.

Zhu, L., Li, L., Xu, L.-H., Cheng, T.-M. 2015. Phase Stability of N Substituted $Li_{2-x} FeSiO_4$ Electrode Material: DFT Calculations. *Comput. Mater. Sci.*, 96, 290–294.

4 Emerging Materials for High-Performance Supercapacitors

Meenu Sharma[1] and Anurag Gaur[2]
[1]Department of Mechanical Engineering, Energy Systems Research Laboratory, Indian Institute of Technology Gandhinagar, 382355, Gujarat, India
[2]Department of Physics, National Institute of Technology, Kurukshetra 136119, India

CONTENTS

4.1 INTRODUCTION

The advent of the industrial revolution at the beginning of the 19th century spurred a large production of goods and services that required an enormous amount of fuel to energize the machinery. Renewable resources of energy have great prospects as future power sources; however, such resources are available only during a particular period or condition. For instance, the sun is not available at night; wind energy can be harnessed only when the wind speed is high. Thus, the intermittent nature of the nonconventional energy sources is a major challenge toward their widespread application and success. The scientific research communities and industries have started working on a clean renewable energy source instead of fossil fuels that cause environmental pollution, global warming, and rapid resource depletion [Bard and Faulkner 2000; Burke 2000]. Efficient and sustainable energy sources hold an important part in the context of energy storage applications. For many technology and commercial purposes, energy storage is the essential technology either to be used directly as energy sources for transport and electronics or to be combined with other energy storage devices, such as thermoelectric and piezoelectric, to efficiently use energy [Zhang et al. 2012; Conway 1999b].

Capacitors and battery cells, for example, are widely used in electrical power systems. To address the issue of intermittency, a charge storage device such as an electrochemical capacitor is employed for energy storage during peak operation of nonconventional resources that can be used at a time of requirement. Electrochemical energy has a lot of advantages including (i) it is clean energy because it prevents any sort of environmental pollution; (ii) it provides a higher power density and efficiency than fuel combustion systems; and (iii) it is abundant and safe [Pandolfo and Hollenkamp 2006]. Generally, the criteria to determine the efficiency of energy storage devices are energy density and power density. Among the charge storage devices, electrochemical capacitors have emerged as a highly significant technology because of their beguiling properties such as simple construction, significant energy density, better cycle life (more than a million cycles), and are comparatively safer than batteries. However, the low efficiency and low practical performance of capacitive materials are the ensuing issues that still need resolution. There is indeed a growing interest in developing the development of nanostructured material having flexibility, transparency, smaller size, higher energy and power densities, and cost-effectiveness for supercapacitors application [Luryi 1988].

In recent years, versatile energy storage has indeed been taken into serious attention with rising wearable devices, including adjustable devices, portable electronics, health-care devices, and so on. In consideration of functional electronics, remotely operated, integrated, and self-powered devices that are assembled in an all-solid-state and can be integrated in the future will be used for long-term concerns with performance and quality. All-solid-state devices can be used without security and production issues for the long term many energy storage systems thus far, electrochemical condensers (or supercapacitors), metal-ion batteries, for example, and most recently, rechargeable metal-air batteries, have been recognized to be the most practical and feasible technologies for all-solid-state energy storage.

However, the main challenges for the current energy storage are still limited by the low volumetric energy density, high internal resistance at the materials interfaces, poor mechanical durability, and controversial environmental concerns [Radhakrishnan et al. 2011; Park et al. 2010].

4.2 SUPERCAPACITORS AND THEIR MECHANISMS

Supercapacitors are also known as ECs or ultracapacitors that have attracted significant attention in energy sustainability due to their excellent electrochemical performance, that is, high specific capacitance, rapid charge/discharge rate, high specific power, and long cycle life ($>10^5$ times) [Park et al. 2010; Conway 1999a], which is incomparable to those of other energy storage devices. Typically, two electrodes, an electrolyte, and a separator are the required components to assemble a full-cell supercapacitor. The configuration of a full-cell supercapacitor can be either symmetric or asymmetric. Two electrodes are separated by a separator (filter paper, glassy paper, cellulose, or polyacrylonitrile membrane), which has good ion permeability properties for ion transportation [Dubal et al. 2015]. According to the energy storage mechanism, supercapacitors are classified into electrical double-layer capacitors (EDLCs) and pseudocapacitors as shown in Figure 4.1.

Pseudocapacitor materials include metal oxides, hydroxides, nitrides, intercalating materials, and polymers. These materials exhibit high capacitance due to the redox reaction. Every molecule participates in the faradaic charge transfer, and thus these materials exhibit high energy density. However, compared to EDLC materials, they exhibit low power density and low-rate capability due to kinetic limitations on the charge transfer reaction. However, pseudocapacitive materials suffer from a low electrochemically active surface area (ESA) that gravely impedes the capacitance obtained from these materials [Graves and Inman, 1965; Conway and Pell 2003]. Pseudocapacitive materials have a high theoretical capacitance ranging from 200 to 3000 Fg^{-1}, owing to their redox behavior. Such high capacitance can lead to high energy density as $E = 1/2CV^2$. However, the practical capacitance or the realizable capacitance remains low; this reduced practical

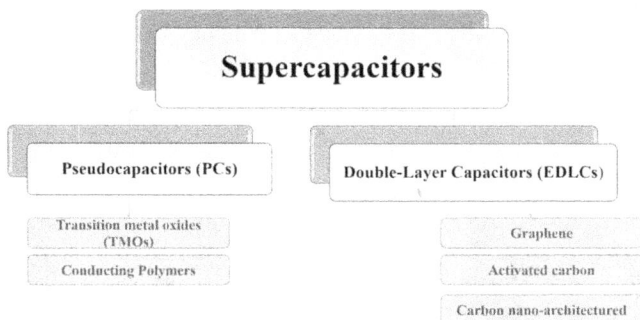

Supercapacitors

Pseudocapacitors (PCs)

Double-Layer Capacitors (EDLCs)

Transition metal oxides (TMOs)

Graphene

Conducting Polymers

Activated carbon

Carbon nano-architectured

FIGURE 4.1 Classification of supercapacitors.

capacitance results from the low and inefficient ESA and high resistance [Silva et al. 2019]. Also, since the electrochemical processes are the surface processes occurring only at the electrode–electrolyte interface, the material inaccessible to the electrolyte remains dead that decreases the gravimetric capacitance of pseudocapacitive material [Radhakrishnan et al. 2011]. Thus, it is necessary to increase the ESA, improve the electrolyte accessibility, and reduce dead volume for increasing or improving the performance of the pseudocapacitive material for wide-scale applications.

Historically, Ewald Georg von Kleist of Pomerania first observed double-layer charge storage in 1745 when he discovered charge storage by connecting a high-voltage electrostatic generator to the flow of liquid by connecting a wire in a glass jar. Since then, the double layer charge storage has progressed a great deal during which glass dielectrics have been replaced with aqueous, nonaqueous, gel, and solid electrolytes, and metal foils are replaced with carbons having high surface area. The new systems store an enormous amount of energy facilitated mostly by high electrode surface, and the high conductivity of materials can be broadly classified into electrochemical double-layer capacitive materials at which charge is stored only at the electrode–electrolyte interface without any transfer of charge and pseudocapacitive materials that perform charge storage through charging transmit and phase change at the electrode–electrolyte interface [Maldonado-Hódar et al. 2000; Zhang and Pan 2015]. Various carbon microstructures with a large effective surface have been used as electrode materials in EDLC. The most widely studied and exploited carbon forms are activated carbons or engineered carbon, carbon nanotubes, carbon nanofiber (CNF), graphene, etc., for the electrochemical capacitor application.

4.3   APPROACHES HAVE BEEN PROPOSED FOR ELECTRODE MATERIALS OF SUPERCAPACITORS

- The electrode materials should have a high areal and volumetric capacity, which represents their ability to supply high energy within a confined space. In addition, some advanced features should be considered in the design of electrode materials such as additive-free and free-standing, which can significantly reduce the content of electrochemically inactive materials in the electrode.
- The electrode materials should have reduced resistance by using high conductivity materials or conducting components toward high-rate performance and less energy consumption by the internal resistance. Experimentally, doping and hybridizing with other elements may improve the conductivity of electrode materials. Computationally, data-driven simulation and machine-learning methods can be used to predict the material compositions and structures to achieve the optimum electrode conductivity. The advanced solid-state electrolyte films with high ionic conductivities should be at nanoscale.
- The ionic conductivities of the electrolyte films are always the barrier for energy storage devices. Similar to the electrode materials, doping may solve the low ionic conductivity issue of the solid electrolyte.

- To design new self-organized nanostructures that enable capabilities to re-lease the interfacial stress with boosting material performance in renewable energy applications. Designing new electrode structures may solve the in-terfacial resistance and stability issues of solid-state devices.
- Environmentally compassionate and nontoxic materials should be utilized. Both electrode materials and solid electrolytes should be very safe and nontoxic to be used for flexible and personal devices. New electrode materials without metal dissolving and leaking should be considered for material design.

4.4 ELECTRODE MATERIALS

The choice and manufacturing process of electrode materials play a key role in improving the capacitive performance of supercapacitors (SCs) [Rajkumar et al. 2015]. SC electrodes shall focus on providing thermal stability, high specific sur-face area (SSA), corrosion resistance, high electrical conductivity, suitable chemical stability, and excellent thermal wettability. They must also be low-cost and eco-logically sustainable. In addition, it is important for their ability to share the electrochemical charge to enhance capacitance performance [Wu et al. 2015]. The specific capacitance aspect is nevertheless affected by surface area; however, there are other key factors, such as simple morphology manipulation, including pore size distributions, pore shapes, pore sizes, and electrolyte availability [Du et al. 2018; Guan et al. 2017]. Consequently, two of the main specifications for the design of SC devices are (i) the improvement of electrochemically active sites by the selection of high SSA electrode materials, and (ii) the arrangement of pore size distribution and pore shapes, e.g. circles, vertical rectangles, horizontal rectangles and squares in the case of graphene nanopores [González et al. 2016], and cylindrical, spherical, and slits in the case of biomass-derivatives. The electrode material having smaller pores offers a higher equivalent series resistance (ESR), therefore, decreasing the power density. On this basis, the application of electrode material affects the selection of materials. For example, for applications where higher peak currents are important, electrode materials should meet the criteria for larger pores. Furthermore, an ap-propriate size distribution can enhance the retention capability, which would be the indication of high power density in the SC device. The efficient system of micro/mesopores of electrode material could provide high-speed mass and ion transport through a continuous pathway, consequently boosting the accessibility of the electrolyte and making the material an appropriate choice for SC applications. Electrode materials are categorized based on three main classes: carbonaceous materials, transition metal oxides (TMOs), and CPs [Sharma et al. 2018]. Figure 4.2 shows the classification of electrode materials for the SC application from a 0D to a 3D structure.

4.4.1 TRANSITION METAL OXIDES (TMOs)

The electric conductivity of metal oxides and their morphologies are considered as one of the important properties that influence the capability of electrochemical supercapacitor (ES) performance. The morphology of metal oxide electrode that

- Solid Nanoparticles
- Hollow Nanoparticles
- Core-shell Nanoparticles

- Nanorods
- Nanowires
- Nanotubes
- Nanopillars

0 D   **1D**

2D   **3D**

- Graphene
- Metal oxise/Hyroxides
- Trasitional Metal Oxides
- Graphene-Metal Oxise/Hyroxides
- Trasitional Metal Carbides/Nitrides

- 3D Carbon Materials
- 3D Pseudocapcitive Materials

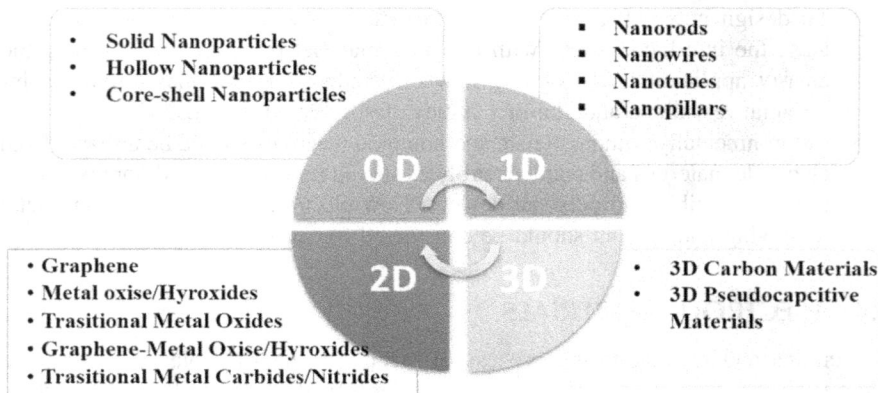

FIGURE 4.2 A schematic representation summarizing the electrode materials from 0D to 3D.

contacts directly with electrolyte must be suitable for collecting huge numbers of charges inside it, such as nanoporous. The various morphologies of the metal oxide electrode exhibit different performances in ESs. The morphologies with 1D material such as nanowires, nanoneedles, nanotubes, etc., are distinguished by the high power density [Chu and Majumdar 2012; Zheng et al. 2017]. Moreover, metal oxides with two or more oxidation states are especially suitable for ES applications owing to increasing reversible redox reactions. TMOs, such as MnO_2, V_2O_5, Co_3O_4, Mn_3O_4, Fe_3O_4, Fe_2O_3, etc., have also shown to exhibit pseudocapacitive charge storage through redox reaction [Ho et al. 2014]. During charge storage, the transition metal atom changes the oxidation state through electron transfer at the electrode–electrolyte interface resulting in faradaic charge storage. Among TMOs, RuO_2 was the first TMO to show pseudocapacitive charge storage and has been extensively studied since then. It possesses quasi-metallic conductivity allowing facile electron transfer, good stability, and involves multiple oxidation state transitions that consequently result in high capacitance under a voltage window of 1.4 V. RuO_2 can show multiple oxidation states, Ru^{2+}, Ru^{3+}, Ru^{4+}, Ru^{6+}, with high reversibility and leads to high specific capacitance. Its high conductivity and presence of structural water in the RuO_2 lattice help in high material utilization exhibiting capacitance as high as 1340 Fg^{-1}, almost 96% of the theoretical capacitance [Wang et al. 2012; Zhang and Pan 2015]. It showed great promise for use in commercial electrochemical capacitors, but it is highly expensive and toxic, which has severely impeded its prospects of application in capacitors. Another TMO that has shown good capacitive behavior is MnO_2: it is low-cost and environment-friendly [Mondal et al. 2015]. Its synthesis is easy and economical and can be prepared from various methods such as electrochemical deposition, hydrothermal syntheses, room temperature reduction of $KMnO_4$, etc. Lee and Goodenough in 1999 demonstrated the charge storage properties of MnO_2 in KCl solution and also showed that strong acids could be replaced by mild aqueous salt solution as electrolyte [Sharma and Gaur 2020a]. In recent years, a novel category of electrode material developed by researchers is mixed (or binary) metal oxides. Recently, research focuses on mixed metal oxides (especially, binary transition metal oxides) because of their superior electron mobility, eco-sustainability, easy fabrication, and high specific capacitance. The binary metal oxides that can be used as active materials for SC applications are $NiCo_2O_4$, $ZnCo_2O_4$, $CuCo_2O_4$, $NiMoO_4$, and $NiWO_4$, etc [Trasatti and Buzzanca 1971]. The advantage of using binary metal oxides is associated with their enhanced electrochemical activity and ultrahigh conductivity as compared to their metal oxide counterparts. The primary reason for such enhanced electrochemical performance is that in binary metal oxides both the individual oxides take part in the redox reaction phenomenon, thus contribute to the overall pseudocapacitance of the fabricated electrode. In the literature, Zhang et al. have synthesized mesoporous $NiCo_2O_4$ nanoneedle morphology, which helps to achieve a reasonable capacitance of 660 Fg^{-1} with 91.8% of cycling stability. While in other studies, Deng et al. have synthesized $NiCo_2O_4$ nanosheets in which sheet-like morphology facilitates better charge transfer that results in increased specific capacitance up to 799 Fg^{-1} using 2 M KOH as an aqueous electrolyte along with good cyclic stability of 87.1% even after 2500 cycles [Sugimoto et al. 2005; Wu et al. 2014]. Xu et al. (2014) worked on mesoporous $NiCo_2O_4$@MnO_2-based core–shell nanowire arrays for aqueous asymmetric supercapacitors (ASCs) as shown in Figure 4.3(a). The charge/discharge curves are shown in Figure 4.3(b), and specific

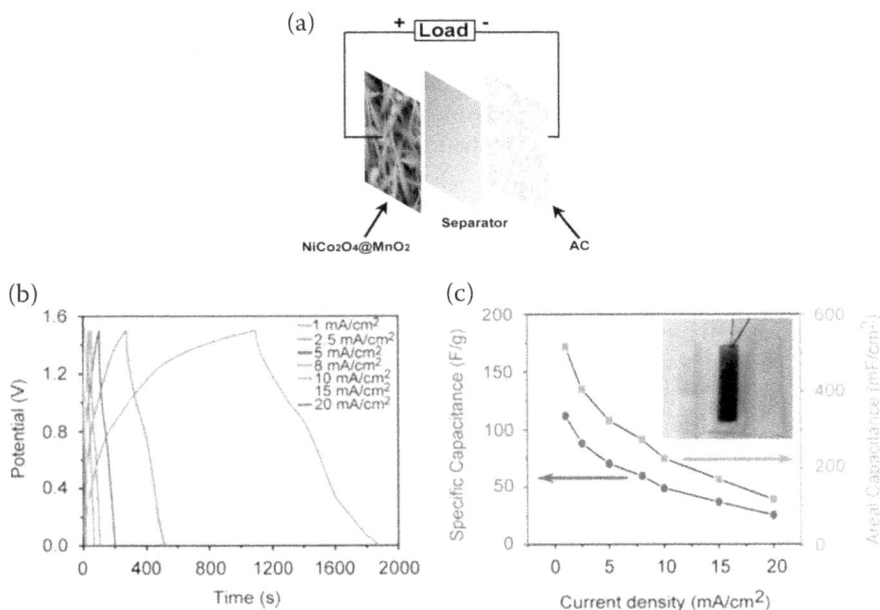

FIGURE 4.3 (a) Schematic illustration of the as-fabricated $NiCo_2O_4$@MnO_2// AC-ASC device based on the hierarchical mesoporous $NiCo_2O_4$@MnO_2 electrode (positive) and the AC electrode (negative). (b and c) Charge/discharge curves and specific capacitances of the $NiCo_2O_4$@MnO_2// AC-ASC device at different current densities. The inset is the digital photograph of the as-fabricated $NiCo_2O_4$@MnO_2// AC-ASC device [Reproduced with permission from Xu et al. 2014].

capacitances of the $NiCo_2O_4@MnO_2//$ AC-ASC device at different current densities, the inset is the digital photograph of the as-fabricated $NiCo_2O_4@MnO_2//$ AC-ASC device, are shown in Figure 4.3(c). The maximum specific capacitance of the device is 112 Fg^{-1} calculated from the charge/discharge curve at 1 $mAcm^{-2}$, which is substantially higher than the previously reported values.

Recently, Huang et al. have synthesized highly porous $NiCo_2O_4$ with a unique strategy via a cotton-assisted template that helps to achieve a high specific capacitance of 1029 Fg^{-1}. However, a very poor cyclic stability of 51.5% was achieved after 2000 charge/discharge cycles. Therefore, a need arises to tune the morphology and synthesis approach via the cotton-assisted template synthesis route to obtain enormous electrochemical performance. Commercially available cotton is a porous structure with a large surface area, and the soft, lightweight, flexible, and low costing nature of the cotton make it ideal for future cost-effective flexible energy storage applications. Further cotton substrate allows the electrolyte to flow throughout the entire surface of electrodes. To proceed with this approach further, active pseudocapacitive materials can be homogeneously deposited into such substrates under fine control of their morphologies and particle sizes. Figure 4.4 represents the plot of energy density and power density for $NiCo_2O_4$-based ASCs.

The $ZnCo_2O_4$-based electrode is considered the most encouraging functional electrode because of its excellent redox reactions, superior theoretical capacitance, eco-goodness of Zn, Co elements, and low cost. Despite possessing low toxicity, low cost, widespread availability, and high theoretical capacitance, it lacks in the low conductivity, low surface area, and rate capability that need to be constantly improved to reach the practical needs of the supercapacitor. Therefore, different architectures of $ZnCo_2O_4$ were studied previously for electrode material of

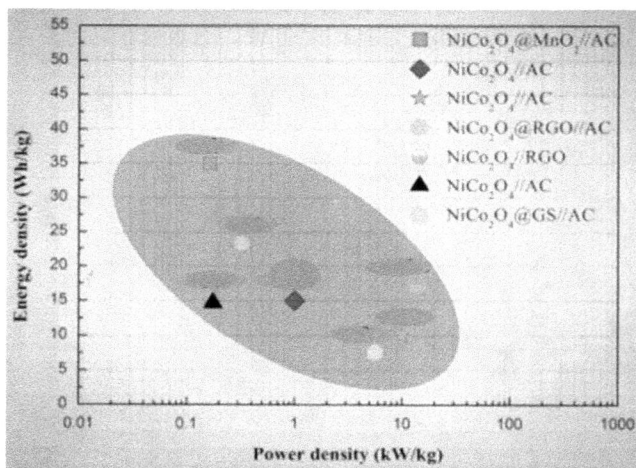

FIGURE 4.4 Figure represents the plot of energy density and power density for $NiCo_2O_4$-based ASCs [Reproduced with permission from Xu et al. 2014 with copyright to RSC 2013; work done by Xu et al. 2014; Lu et al. 2014; Hsu and Hu 2013; Wang et al. 2012; Chen et al. 2014; Ding et al. 2013; Wang et al. 2012].

TABLE 4.1

Supercapacitors Materials Based on Spinel Ternary Metal Oxides

Different metal oxides and combinations	Synthesis method	Capacitance	References
$NiCo_2O_4/MnO_2$	Hydrothermal	5.3 Fcm^{-2}	Su et al. 2018
$ZnCo_2O_4/Ni\ (OH)_2$	Electrochemical deposition	4.6 Fcm^{-2}	Su et al. 2018; Pan et al. 2017
2D-$LiCoO_2$	Electrochemical deposition	310 $mFcm^{-2}$	Su et al. 2018; Lu et al. 2020
$CuCo_2O_4$	Solution combustion	338 Fg^{-1}	Su et al. 2018; Jadhav et al. 2016
$MnCo_2O_4$	Hydrothermal	349.8 Fg^{-1}	Su et al. 2018; Kumbhar et al. 2018
$MgCo_2O_4$	Molten salt method	321 Fg^{-1}	Bao et al. 2011
MnO_2/MoS_2	Magnetron sputtering	224 $mFcm^{-2}$	Zhang et al. 2020
$CoMn_2O_4$	Solvothermal	321 Fg^{-1}	Gao et al. 2019
$ZnCo_2O_4$	Hydrothermal	290.5 Fg^{-1}	Al Haj et al. 2019

supercapacitor. The electrode efficiency can be improved by changing the inherent conductivity through doping of other transition metal ions. The electrochemical performance of electrode material could be improved by changing the inherent conductivity through doping of other transition metal ions. Several reviews are already applied to the development and manufacturing of supercapacitors based on $ZnCo_2O_4$ composites with carbon-based materials and different types of polymers [Yuan et al. 2009]. Thus, the merging of electrode materials with different characteristics in a single composite with enhanced electrochemical properties is a possible approach to resolve such issues. Table 4.1 summarizes some reported supercapacitors materials based on spinel ternary metal oxides.

4.4.2 CARBON NANOSTRUCTURED BASED ELECTRODE MATERIALS

Carbon-based materials are being used to develop electrodes for EDLC devices to enhance their capacitance performance. The inherent properties of electrode materials such as SSA, distribution of pores, and pore size extremely affect the capacitive performance of an EDLC device. Consequently, owing to the large SSA and the high porosity of the carbonaceous materials, they have displayed enhanced capacitance.

In addition, carbon-based materials are known for their high electrical conductivity, chemical, thermal, and electrochemical stability (in various solutions from acidic to basic). Moreover, the carbon-based materials exhibit symmetrically high Galvanostatic charge/discharge profile and a good rectangular profile of cyclic voltammetry (CV) curves [Iro 2016]. These characteristics of carbon-based materials manifest their usefulness as capacitive materials. The physicochemical characteristics of the carbon-based materials depend on the pore size that is approximately less than 1 nm [Chen et al. 2014]. The charge storage mechanism of the carbon-based electrode mainly depends on the formation of a thin layer on the surface of electrode material known as the Helmholtz layer, which is in contact with the electrolyte ions (both nondissolved and very weakly dissolved ions). The charge storage derives from the formation of a thicker outer Helmholtz layer, which is made through the existence of the solvated ions conducted by strong electrostatic forces [Filleter et al. 2011].

Carbon-based materials can be used in various forms such as powder, fiber, and foils in an SC device. There are several common examples of carbon-based materials such as carbon aerogels, graphene, AC, activated carbon bras (ACF), carbon nanotube (CNT), and carbon cloth (as well as various carbon-based composites), which are suitable to be used as the electrode materials in SC devices and are explained in detail in the following sections [Li et al. 2007]. However, there is a huge necessity to enhance the capacitance of the carbon-based material to obtain high energy density.

4.4.2.1 Graphene

Graphene is defined as "a single carbon layer of graphite structure, describing its nature by analogy to a polycyclic aromatic hydrocarbon of quasi-infinite size." Graphene is an unusual material that has far-reaching applications encompassing the fields of medicine, material science, engineering, nanotechnology, energy, etc. After its discovery, a significant number of studies on graphene have been published, indicating the scale of work focusing on graphene. Graphene consists of highly delocalized electrons called pi-electron; these dislocated electrons help in providing graphene with exceptional electronic and thermal conductivity [Bonaccorso et al. 2012]. High-quality graphene can be synthesized by molecular beam epitaxy, micromechanical cleavage, photoexfoliation, precipitation from metal, etc. However, these methods produce graphene in low quantities. For large-scale graphene synthesis, Hummer's method and exfoliation of graphite in organic solvents are well-known methods. The characteristics such as high SSA, conductivity, and efficiency make it the best material for electrochemical capacitors. Also, graphene-based composites with transitional metal oxides have shown improved electrochemical properties.

Properties like high SSA (2630 m^2g^{-1}), good electrical conductivity, flexibility, and high mechanical strength make it an ideal material for electrochemical capacitors with the theoretical capacitance of 550 Fg^{-1}. Rao et al. demonstrated the application of graphene in EDLC in 2008 and later by Ruff et al. Gravimetric capacitance obtained for graphene ranges from 150 to 298 Fg^{-1} depending upon the synthesis method, surface area, pore size, etc [Chen et al. 2008; Marcano et al. 2010]. However, the capacitance of graphene is limited by the quantum capacitance phenomenon caused by the low density of states that put the limit of 13.5 μFcm^{-2}. Heteroatom doping of graphene with nitrogen, boron, phosphorus, sulfur, etc. has been shown to improve the capacitance of graphene. For nitrogen-doped single-layer graphene, the areal capacitance has been shown to improve to 23 μFcm^{-2} compared to 13.5 μFcm^{-2} for pristine graphene Wang et al. 2011].

4.4.2.2 Graphitic Carbon Nitride

Among the various carbon-based nanostructures, graphitic carbon nitride (g-C_3N_4) is a soft polymer with porous nature that has attracted considerable attention because of its highly active nitrogen sites, excellent physical and chemical strength, and low-cost feature [Fazal-ur-Rehman 2018]. In the application of water splitting, wastewater detoxification, solar cells, and supercapacitors, carbon-based materials (g-C_3N_4) show superior performance because of their outstanding optical properties, high mechanical strength, and thermal conductivity [Sharma and Gaur 2020b]. The nitrogen presence in g-C_3N_4 provides more active sites and advances the capacitance while preserving the cyclability of the electrochemical device. To date, only a few studies of g-C_3N_4 with Ni(OH)$_2$, MnO_2, and $NiCo_2O_4$ for energy storage application have been reported [Wu et al. 2020; Li et al. 2017; Chang et al. 2017]. The composites of g-C_3N_4 and $ZnCo_2O_4$ could have great advancement in rate capability and specific capacitance. In a study, Sharma and Gaur (2020b) synthesize a g-C_3N_4-hybridized $ZnCo_2O_4$ composite for the electrode material to develop a high-performance symmetric supercomputer device. This hybrid g-C_3N_4@$ZnCo_2O_4$ composite exhibits excellent electrochemical performance through a specific capacity of 154 mAhg^{-1} at 4 Ag^{-1} with 90% of capacity retention up to 2500 cycles as shown in Figure 4.5(b) & (c).

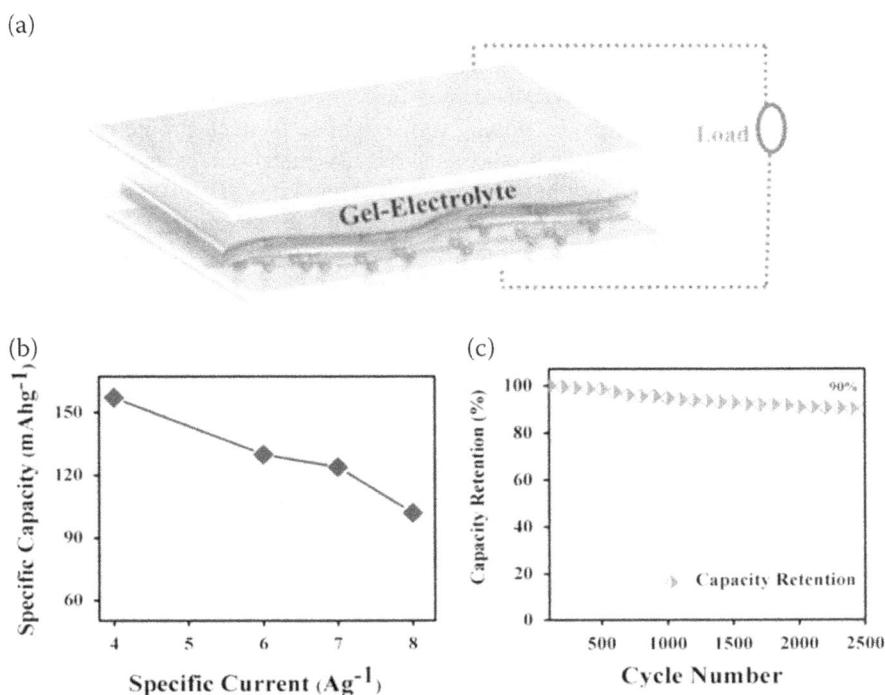

FIGURE 4.5 (a) Schematic illustration of the assembled g-C_3N_4@$ZnCo_2O_4$//g-C_3N_4@$ZnCo_2O_4$ symmetric device using gel electrolyte. (b) Specific capacity as a function of specific current for g-C_3N_4@$ZnCo_2O_4$ sample. (c) Capacity retention up to 2500 cycles for g-C_3N_4@$ZnCo_2O_4$ sample [Reproduced with permission from Sharma and Gaur 2020b].

4.4.2.3 Carbon Nanotubes (CNTs)

CNTs, discovered in 1991 by Sumio Iijima, are cylindrical nanostructures of carbon. They have a tubular structure; the properties of CNTs depend on the axis of rolling. The pristine CNTs are made up of an sp^2 hybridized carbon layer that is rolled into tubes. CNTs possess very unusual properties, which make them an important material in electronics, optics, and material science. CNTs are the stiffest materials discovered to date with a tensile strength of 63 giga pascal and specific strength of 8000 kNkg^{-1} (high carbon steel has specific strength of 154 kNm kg^{-1}). From the point of view of electrochemical applications, it possesses high electrical conductivity in the range of 10^4–10^5 Scm^{-1} along with high thermal conductivity (3500 Wm^{-1}K^{-1}), which provides excellent heat dissipation during device operation [Peigney et al. 2001]. This extraordinary electrical and thermal conduction is caused by the nanoscale cross-section that allows conduction only along the tube axis giving rise to ballistic transport [Niu et al. 1997]. Depending upon the number of carbon layers, they are classified as single-walled carbon nanotubes (SWCNTs), double-walled carbon nanotubes (DWCNTs), and multi-walled carbon nanotubes (MWCNTs). CNTs also possess high SSA, which is 1315 m^{2}g^{-1} for SWCNT and 300 m^{2}g^{-1} for MWCNT and is important for electrochemical applications. CNTs were first applied in electrochemical capacitors by Niu et al. (1997). MWCNTs with a surface area of 430 m^{2}g^{-1} showed the capacitance of 102 Fg^{-1}. Since then CNTs have been extensively researched for electrochemical capacitors [Liu et al. 2020]. CNTs have been modified through surface functionalization using chemical and physical approaches, heteroatom doping to manipulate their electronic properties, making composites of CNTs with other material like metal oxides and conducting polymers (CPs). CNTs have a high surface-to-volume ratio that makes them ideal for the deposition of other capacitive material along the axis where the CNT core can provide better current collection and transport.

4.4.2.4 Activated Carbon

Activated carbon with microporous (pore size: <2 nm) and mesoporous (2–50 nm) structure is the most widely used material in supercapacitors. It is produced from carbonaceous materials like coconut husk, bamboo, cotton, and coal. During synthesis, carbonaceous materials are first pyrolyzed at 450–900 °C in inert gases such as nitrogen or argon to obtain the carbon. This carbon is then activated mainly by two processes: physical activation and chemical activation. When the carbon is physically activated, it is exposed to oxidizing atmospheres such as steam or oxygen at temperatures between 600 and 1200 °C. Activated carbon possesses a high SSA up to the order of 3000 m^{2}g^{-1}. Thus, it can store an enormous amount of charge at the interface compared to the conventional metal capacitors. Also, it possesses high electronic conductivity that results in high power capability, making it ideal for high power applications. As the charge storage is essentially taking place at the surface, the process is highly reversible with high response time, and as the chemical nature of carbon is not affected within safe working voltages, they usually exhibit high cycle life (can be cycled millions of times) [Yang et al. 2018]. In addition to the above advantages, carbon can be operated in a potential range of 2.5–3.0 V in organic electrolytes. Thus, a large operating voltage and high energy density could be achieved. The charge storage of activated carbon can be further improved by introducing surface functional groups that provide significant pseudocapacitance to the carbon.

4.4.2.5 Carbon Nanofibers

The layers of graphene in the arrangement of plates give rise to cylindrical nanostructures called CNFs. The earliest technical records for the CNFs date back to 1889 by Hughes and Chamber for the filamentous carbon. CNFs are easy to synthesize, have a high aspect ratio, and the capability to form free-standing films make them interesting for many electrochemical applications such as supercapacitors, batteries, fuel cells, etc. CNFs are prepared through chemical vapor deposition technique in which gas-phase molecules (hydrocarbons) are decomposed over metal such as MgO, Al_2O_3, Mn, Zn, etc. The decomposition of hydrocarbons results in the deposition of carbon over the catalyst around which the growth of carbon fiber takes place. Versatile properties of CNF have led to its application in rechargeable Li-ion batteries, as a field emission source, and also in electrochemical capacitors. A relatively different technique involving the synthesis of polyacrylonitrile films through electrospinning followed by stabilization and carbonization produces continuous CNFs.

4.4.3 Conducting Polymer (CP)-Based Material

CPs are just another pseudocapacitive electrode material widely studied due to their high capacitance, chemical modification adjustable redox activity, strong doped state conductivity, high voltage windows as well as easy processing, low environmental effect, and low cost. As no structural alterations including phase changes arise during the charge/discharge mechanism, CPs could store the charge in its bulk. Because of this, CPs can have greater capacitance due to wider surface areas and redox storage capabilities [Cheng et al. 2011]. CPs have higher conductivity, capacitance, and low ESR as compared to the other carbon-based electrode materials. In CPs, the specific capacitance is extracted from the fast-reversible redox reactions where the ions pass through the oxidation phase, often defined as doping, to the polymer backbone and released back into the electrolyte solution during the reduction processor-doping process.

Nevertheless, considering all the benefits, mechanical stress in CPs by reduction-oxidation limits the stability of electrodes during charge/discharge cycles Sharma et al. 2008]. Also, due to the slow ion diffusion rates in bulk, efficiency and power densities get hindered, which are the major disadvantages of CPs. Further, polymer hydrogels are also successful candidates for SC electrodes. To synthesize the polymer hydrogel, various strategies, such as (i) in situ monomer polymerization and (ii) the introduction of cross-linkers through the polymerization process are used for the process of gelation. The first method, however, often offers favorable mechanical performance; the electrochemical performance is not satisfactory even though it offers favorable mechanical performance whereas the second one showed good electrochemical behavior. Owing to their resistance to deformation and physical damage, CP hydrogels may also be used as electrolytes in SCs [Sivaraman et al. 2006]. In a study, an SC was developed inside the two electrodes by inserting the PEI-PVA-Bn-LiCl hydrogel electrolyte, which showed superior specific capacitance with the extended operating potential window, as well as high cycling stability and superior mechanical stability.

4.4.3.1 Polyaniline (PANI)

Polyaniline (PANI) is among the most common polymer electrode material to conduct due to its high conductivity, easy fabrication, outstanding energy storage capacity, excellent stability, and low cost. In addition, PANI is mechanically versatile and eco-friendly. In addition to these advantages, PANI is prone to severe practical application because of the degradation of performance due to repetitive charging/discharge. To fix this concern, the pairing of PANI and carbon materials was used as a PANI layer and coated on metal oxides (MOs)/carbon composite to develop a hybrid composite (PANI/MOs/carbon), which provided enhanced cyclic stability for PANI and improved capacitance. In addition, PANI exhibits a wide variety of electrochromic properties because of its various oxidation and protonation forms. Such character traits consider it a great choice for the manufacture of electrochromic SCs [Luo et al. 2013]. PANI has been used as an active SC material using an electrochemical polymerization method by Wang et al. with an array in a unique nanowire structure that leads to a reduction in diffusion pathways, as well as a load transfer resistance, which allows the material to have a high specific capacitance even at high current densities.

4.4.3.2 Polypyrrole (PPy)

The necessity to establish energy storage technologies for rising electrode materials could be extracted from the structurally controlled conductive polymer hydrogels with mechanical flexibility and adjustable electrochemical properties. Polypyrrole (PPy) is just one of those polymeric materials that draw much more attention from researchers. This has a greater density and even more flexibility compared to other PCs. This could endure a rapid redox reaction for charging storage, as well as a high electrical conductivity value around 10 and 500 Scm^{-1}. The conductive hydrogel nanostructure has been synthesized using an interfacial polymerization technique and accessed a 3D porous PPy network that showed excellent rate and high specific capacity. In a study, Zhang et al. (2013) worked on PPy with graphene hydrogel (GH) nanocomposites for supercapacitor applications. Figure 4.6(a) shows the Galvanostatic charge/discharge curves of PPy at different current densities, (b) galvanostatic charge/discharge curves of PPy/GH15 at different current densities, (c) specific capacitance of PPy and PPy/GH15 varying with current density, and (d) a Ragone plot for all existing energy storage systems for comparison.

The synergistic effect of the pseudocapacitance of PPy and the electric double-layer capacitance of GH, PPy/GH15 exhibits a high specific capacitance of 375 Fg^{-1} at a scan rate of 10 mVs^{-1}. In addition, the PPy/GH15 nanocomposite electrode has a much higher energy density and power density than traditional capacitors and industrial supercapacitors, indicating that it may be a promising electrode material for high-performance supercapacitors (Figure 4.6).

(a)

(b)

(c)

(d)

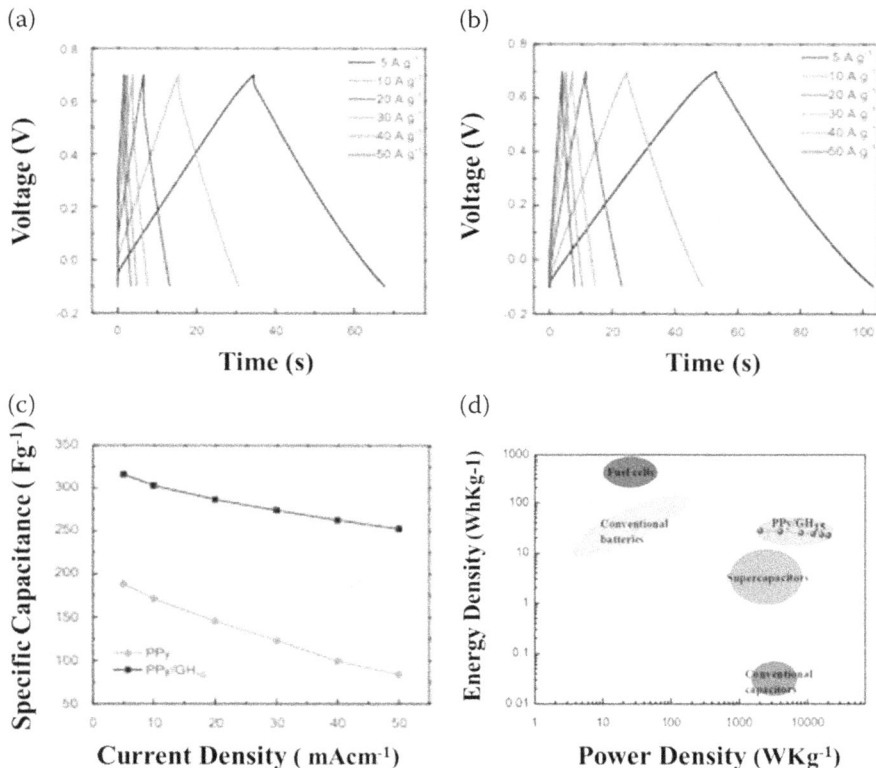

FIGURE 4.6 (a) Galvanostatic charge/discharge curves of PPy at different current densities. (b) Galvanostatic charge/discharge curves of PPy/GH15 at different current densities. (c) Specific capacitance of PPy and PPy/GH15 varying with current density. (d) A Ragone plot was added to the energy density–power density map for all existing energy storage systems for comparison [Reproduced with permission from Zhang et al. 2013].

4.4.4 NANOCOMPOSITE-BASED MATERIALS: THE ROADMAP TO THE HYBRID MATERIAL

Electrodes based on nanocomposites have combined carbon-based materials with either CP or metal oxide intending to have synergetic properties in a single electrode. There are several types of nanocomposites for SC electrodes such as carbon–metal oxides composites, carbon–CP-based material composites, carbon–carbon compounds, and metal oxide–CP-based materials resulting in a notable increment in specific capacitance whereas the SSA is strengthened by the existence of carbonaceous nanocomposite-based electrode materials.

4.4.4.1 Carbon–Metal Oxide Composites

Nanostructures of carbon (nanoparticles, nanotubes, sheets, and porous 3d architectures) are known to be composited with metal oxides to improve the power and energy densities of electrode materials. A typical carbon-based supercapacitor involves symmetrical activated carbon electrodes, where capacitive charge storage happens in the electrochemical double layer. To invoke more complex designs and to improve charge storage, redox-active and pseudocapacitive materials such as oxides are used as an alternative to carbon electrodes. However, in these oxides and redox-active materials, increased capacitance collaterally brings a reduced cycle life and power density. Hence, a concept design is developed in the past few years to composite the best features of two worlds, where carbon-based charge capacitance meets oxide-based faradic capacitance in the form of a matrix of the former.

While oxides, in general, are used as an electrode in batteries, their composite with activated carbon electrodes hybridizes the electrode design resulting in better cycle life and capacitance as well. Further, in some specific cases such as CNT-MnO_2 composites, high mechanical stability also adds to the list of improvements owing to compositing these materials from two domains. Similar to CNT, composites of graphene with MnO_2 are also known to exhibit high capacitance and high cycle-life ($>10^4$). Other examples of composites with carbon black and mesoporous carbon [Lei et al. 2008] are also known to exhibit better performance in terms of capacitance and cycling rates. In various other examples, reduced graphene oxide (rGO) is hybridized with oxides of iron and vanadium, Fe_3O_4 and V_2O_5. In the sheet matrix or CNT backbone, integration of carbon results in a better performance of these oxides as supercapacitors because the shortcomings like poor conductivity are overcome in composites. Apart from CNT and RGO, carbon nanofoams are also used with iron oxide coatings to achieve augmented charge/discharge capacity and faradic capacitance. In the case of V_2O_5, various examples of carbon composites are known, among which an improvement of up to 20,000 charge/discharge cycles with 88% capacity retention is reported, recently [Ngom et al. 2020]. In the first row of transition metals, other oxides like NiO and mixed oxides like $NiMoO_4$ are known to exhibit excellent performance and improvement in their integration with various forms of the carbon matrix.

Similar to the first row of transition metals, other transition metal oxides like RuO_2 also exhibit the improvement in both energy and power densities in their composite form and thus have been scrutinized heavily for supercapacitor applications. The atomic layer deposition method can also be used to develop a CNT version of composites of RuO_2, which leads to an enhancement in mechanical stability, power density, and an enhanced number of charge/discharge cycles (10^4) [Warren et al. 2015]. In summary, a carbon–metal oxide composite is a balancing act of material and electrochemical properties, where carbon provides a highly stable and conductive framework, and metal oxide provides faradic capacitance and sophisticated features of pseudocapacitance. Their immersion in each other brings the best out of both domains, where high capacitance, the longevity of charge/discharge cycles, mechanical stability, and high power and energy densities are the encouraging improvements seen based on various combinations. In addition, various allotropic forms of carbon also complement variation factors like morphological and fabrication level improvements.

4.4.4.2 Carbon–Carbon Composites

The high capacitance of carbon material results from the effective SSA that electrolyte ions can achieve. As a consequence, increasing the SSA of carbon-based materials leads to higher energy in SCs. Different properties of carbon materials that could be used to create carbon-based composites have been identified. For example, graphene nanostructures were synthesized and reported that the material tends to be aggregated due to the presence of van der Waal forces through the chemical reduction process in graphene oxides, leading to a less accessible surface area of electrolyte ion. However, great results were seen from the graphene aerogel SC.

4.4.4.3 MO/CP Composites

The MOs/CPs-based material is another alternative composite for electrode material that provides an enhanced electrochemical characteristic in SCs resulting from the compatibility between the MOs and CPs. The specific capacitance, rate capability, and cyclic stability can be increased, in comparison with pure MOs and pure CPs material for electrodes, by developing the composites of MOs coupled with CPs, which leads to improved conductivity of the electrodes. For instance, Sari et al. (2019) worked on an ASC that consists of the negative electrode that is a novel 3D networked PPy NT/N-doped graphene (NDG) and the positive electrode that is a core-shelled MoO_3/PPy-supported MoS_2. Figure 4.7(a) shows the schematic diagram of the formation of PPy nanotube/NDG. The CV and specific capacitance of ASC at different scan rates are measured as shown in Figure 4.7(b and c). The ASC also shows excellent electrochemical stability with capacitive retention of 126% after 5000 cycles (Figure 4.7d), demonstrating its potential application for next-generation energy storage devices. The remarkable electrochemical performance of individual electrodes is also presented (Table 4.2).

4.5 AN OUTLOOK FOR 2D ELECTRODE MATERIALS

The 2D material with a layered structure has drawn much attention from scientists because of its incredible properties since the enormous discovery of exfoliated monocrystalline graphitic films in 2004. Various 2D nanomaterials were obtained by various preparations and exfoliation for 16 years after the effective mechanical exfoliation of graphene from 3D graphite. Different 2D nanomaterials such as hexagonal boron nitride, metal nitrides, TMOs, black phosphorus, and transition metal dichalcogenides have been isolated from various layered precursor compounds like TMDs (transition metal dichalcogenides), metal-organic structures (metal-organic frameworks, MOFs), and graphitic carbon nitride (g-C_3N_4) [Poonam et al. 2019]. By changing the electrode materials and altering the substance of the electrolyte, the efficiency of the SC device could be improved primarily. Further, various modification approaches can provide more important insights into the study of 2D materials in addition to their special intrinsic properties. There are several standard modification techniques to increase the surface area that contribute to improving the charge storage potential of the electrode materials, thereby improving the properties of 2D materials. It is possible to divide these approaches into several groups, such as nanosizing hybridization and intercalation.

Furthermore, to achieve a high power/energy density SC device, advanced electrode materials are needed to have acceptable properties based on the individual application. There are various ways to enhancing the electrochemical properties of the

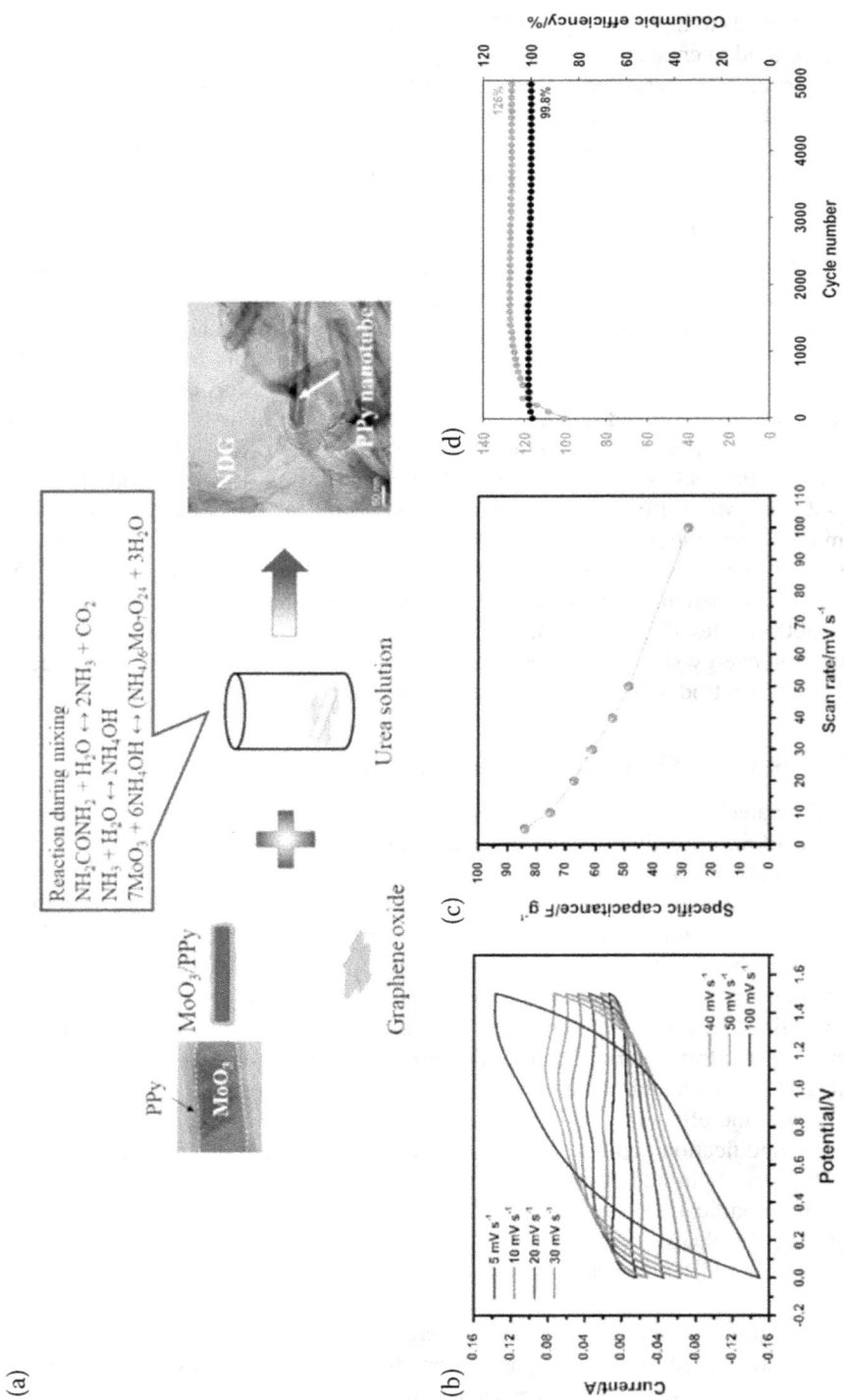

FIGURE 4.7 (a) Schematic showing the formation of PPy nanotube/NDG. (b) CV curves and (c) specific capacitance of ASC at different scan rates. (d) Cyclic stability after 5000 cycles at 200 mVs⁻¹ of ASC [Reproduced with permission from Sari et al. 2019].

TABLE 4.2
Graphene- and CP-based Composites

Electrode material	Electrolyte (M)	Capacitance (F/g)	Current density	Retention cycle	References
Graphene/PPy	LiClO₄ (0.1)	1510	—	—	Mini et al. 2011
GNS/PANI	H₂SO₄ (1)	1130	—	1000	Li et al. 2011
Graphene/PANI	H₂SO₄ (1)	1126	—	1000	Wang et al. 2010a
rGO/PANI	H₂SO₄ (0.5)	970	2.5 A/g	1700	Xue et al. 2012
PANI/Graphene	HClO₄ (1)	878	1.0 A/g	1000	Hu et al. 2012
GO/PANI	H₂SO₄ (1)	746	0.2 A/g	500	Wang et al. 2010b
Graphene/PANI	H₂SO₄ (1)	640	0.1 A/g	1000	Feng et al. 2011
Graphene/PANI	H₂SO₄ (2)	526	0.2 A/g	—	Mao et al. 2012
G-doped PANI	H₂SO₄ (1)	531	0.2 A/g	—	Wang et al. 2009

pseudocapacitive materials such as MnO_2 and RuO_2 to attain advanced pseudocapacitive materials such as TMOs and metal-organic frameworks (MOFs). Wang et al. (2015) tested the performance of MOF-based solid-state SCs for flexible and wearable electronics by sandwiched two identical pieces of PANI-ZIF67-CC electrodes using gel electrolyte [Figure 4.8(a)]. The electrical performance of the device did not deteriorate even after bending and twisting [Figure 4.8(b)–(d)]. We evaluated the capability of using these SSC devices to power small electronic devices, such as a light-emitting diode (LED). Three SSCs were connected in series and charged by two 14500 batteries in series for 30 s, and a red LED was successfully lighted up by these SCs [Figure 4.8(e)]. Figure 4.8(b) represents the maximum areal capacitance of 2146, 1466, and 901 $mFcm^{-2}$ for PANI-ZIF-67-CC electrodes at the scan rates of 10, 50, and 100 mVs^{-1}, respectively, which is more than other reported electrodes.

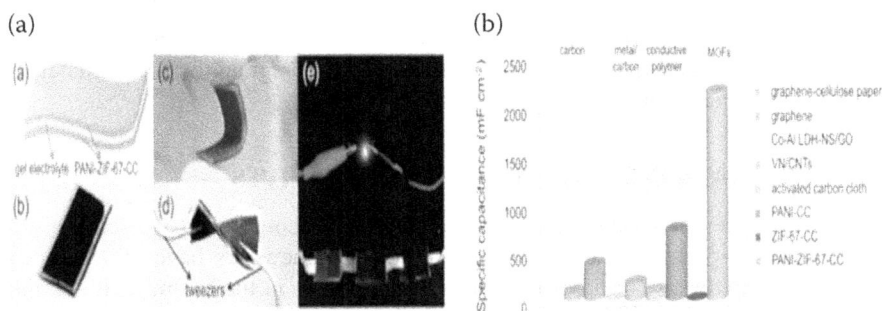

FIGURE 4.8 (a) Schematic illustration of PANI-ZIF-67-CC flexible solid-state SC device. (b–d) Optical photographs of the fabricated flexible solid-state SC device under (b) normal, (c) bent, and (d) twisted states. (e) Photograph of a red LED powered by the three SCs connected in series. (b). Areal capacitances of the literature reported supercapacitors, and PANI-ZIF-67-CC are presented [Reproduced with permission from Wang et al. 2015].

For instance, the hybridization of 2D-TMDs with different carbon-based supporting materials has been documented in several studies to increase more active sites and ease the transferability of architecture. The perovskite (PSK) material is one of the advanced substitute materials that have attracted a great deal of interest as SC electrode materials. As the anode material for SCs to have an anion intercalation mechanism, PSK could be used. Because of their improved physicochemical properties and the controllable oxygen vacancy number, PSK oxides are promising multi-functional materials, which could be tuned by different techniques to improve the properties of the material of PSK oxides as anodes for anion-intercalation SC [Wang et al. 1999]. The proportion of oxygen vacancies and the high SSA of PSK oxides could improve the electrochemical performance of SCs. PSK materials, such as lightweight PSK solar cells and flexible supercapacitors (FSCs), are also good candidates for flexible energy storage/conversion.

4.6   AN APPROACH TOWARD STRETCHABLE ELECTRODES

Stretchable electrodes are those flexible electrodes that can maintain the function under extreme conditions, such as large deformation. They can be wrapped conformally around complex and unconventional shapes including body shapes. Their use of wearable electronics can minimize discomfort. Stretchable electrodes can be achieved in two ways: the use of new structural layouts in conventional materials and new materials in conventional layouts [He et al. 2013]. The former strategy is commonly used. The structural layouts to accommodate the applied strain mainly include wavy structure, net-shaped structure, and helical shape, thus avoiding substantial strain on the material itself. At present, stretchable electrodes for batteries/supercapacitors are fabricated by integrating an active material onto/into an elastic substrate that is either a polymer or a textile. Elastic polymer substrates include polydimethylsiloxane (PDMS) [Lu and Xia 2006], latex, and poly (styrene-block-isobutylene-block-styrene) (SIBS). The elastic or stretchable fabric is the ideal substrate for those electrodes that can be made into breathable textile formats with stretchability. In addition, the intrinsic porous structure of textile is beneficial to electrolyte and ion access leading to better electrochemical. Similar techniques as those for planar and fiber substrates can be applied to integrate the active materials with the elastomeric substrates.

4.7   APPLICATIONS AND FUTURE PERSPECTIVES

Many types of research on the advancement of electrode materials, the study of the supercapacitors process, and its charging storage frameworks have allowed new perspectives on the use of advanced electrode materials in the asymmetric flexible hybrid supercapacitor increasingly in forthcoming. The specification of the hybrid-asymmetric superconductor will provide them access to a wider operating potential window resulting from the unique potential range of cathodes as well as anodes. Hybrid supercapacitors, however, characterize the benefits of both the battery and the supercapacitor. The electrodes are designed to provide high energy density and power density, respectively [Miller 2013].

FSCs have controlled a lot of effort as one of the most important energy storage research centers due to their extraordinary capacitive behavior, high power density, high capacity, and high cyclic stability compared to batteries and conventional capacitors, which provide them with a wide range of potential to be used in various energy storage applications. However, there are some other obstacles to overcome concerning the manufacturing of such devices, along with cost-effective and scalable manufacturing methods, and the development of active strategies to provide more influence over the surface topography (e.g. porosity, morphology, etc.) of the materials.

Almost every classification of SC seems to have its unique features and functionalities, e.g. in portable/wearable electronic devices as well as in computers and mobile phones [Habibzadeh et al. 2017], high power sources for heavy lift trucks or cranes, self-powered SCs for medical applications (e.g. flexible artificial skin, artificial intelligence, portable or implantable medical/biomedical devices, and solar-powered SCs for medical applications). The most intense concern in the development and manufacturing of FSCs to accomplish sufficient mechanical strength and high capacitance utilizing unlimited, nonbinding electrodes is the choice of electrode materials that are the major features of FSCs. The criteria of remarkable electrical conductivity, high tensile strain, high SSA, favorable chemical and thermal stability, high flexibility, large potential windows, and large functional surface groups must be fulfilled by electrode material for flexible supercapacitors.

Among carbon-based materials, for example, graphene is among the strongest candidate for electrode material in FSCs due to a large SSA and high flexibility, which further provide graphene the power to twist to high angles and provide a wide range of microstructures

Aerogels have recently demonstrated superior applications for 3D-printed energy storage applications. Graphene aerogels, for example, are widely used due to their high SSA, mechanical properties as well as tuneable pore structures, and porosity. Consequently, as published in different articles Zhang et al. 2020], carbonaceous materials might be used as a foundation in composite materials also with redox-active materials such as metal oxides and CPs or could have been implemented with novel 2D materials like flexible SCs. In addition, the incorporation of redox-active small molecules and bioderived functional groups into pseudocapacitive materials, as well as their use to increase the capacitance of carbon-based materials, have been provided with more efficient, low-cost, nontoxic SC-based electrodes through eco-friendly synthesis.

REFERENCES

Al Haj, Yazan, Balamurugan Jayaraman, Nam Hoon Kim, and Joong Hee Lee. 2019. "Nitrogen-Doped Graphene Encapsulated Cobalt Iron Sulfide as an Advanced Electrode for High-Performance Asymmetric Supercapacitors." *Journal of Materials Chemistry A*, 7(8), 3941–3952. doi:10.1039/C8TA12396A.

Bao, Lihong, Jianfeng Zang, and Xiaodong Li. 2011. "Flexible Zn_2SnO_4/MnO_2 Core/Shell Nanocable–Carbon Microfiber Hybrid Composites for High-Performance Supercapacitor Electrodes." *Nano Letters*, 11(3), 1215–1220. doi:10.1021/nl104205s.

Bard, Allen J., and Larry R. Faulkner. 2000. *Electrochemical Methods: Fundamentals and Applications*. 2nd ed. Wiley.

Bonaccorso, Francesco, Antonio Lombardo, Tawfique Hasan, Zhipei Sun, Luigi Colombo, and Andrea C. Ferrari. 2012. "Production and Processing of Graphene and 2d Crystals." *Materials Today*, 15(12), 564–589. doi:10.1016/S1369-7021(13)70014-2.

Burke, Andrew. 2000. "Ultracapacitors: Why, How, and Where Is the Technology." *Journal of Power Sources*, 91(1), 37–50. doi:10.1016/S0378-7753(00)00485-7.

Chang, Xueting, Xinxin Zhai, Shibin Sun, Danxia Gu, Lihua Dong, Yansheng Yin, and Yanqiu Zhu. 2017. "MnO_2/g-C_3N_4 Nanocomposite with Highly Enhanced Supercapacitor Performance." *Nanotechnology*, 28(13), 135705. doi:10.1088/1361-6528/aa6107.

Chen, Haichao, Jianjun Jiang, Li Zhang, Tong Qi, Dandan Xia, and Houzhao Wan. 2014. "Facilely Synthesized Porous $NiCo_2O_4$ Flowerlike Nanostructure for High-Rate Supercapacitors." *Journal of Power Sources*, 248(February), 28–36. doi:10.1016/j.jpowsour.2013.09.053.

Chen, Jian-Hao, Chaun Jang, Shudong Xiao, Masa Ishigami, and Michael S. Fuhrer. 2008. "Intrinsic and Extrinsic Performance Limits of Graphene Devices on SiO_2." *Nature Nanotechnology*, 3(4), 206–209. doi:10.1038/nnano.2008.58.

Chen, Shen-Ming, Rasu Ramachandran, Veerappan Mani, and Ramiah Saraswathi. 2014. "Recent Advancements in Electrode Materials for the High-Performance Electrochemical Supercapacitors: A Review." *International Journal of Electrochemical Science*, 9, 4072–4085.

Cheng, Qian, Jie Tang, Jun Ma, Han Zhang, Norio Shinya, and Lu-Chang Qin. 2011. "Graphene and Nanostructured MnO_2 Composite Electrodes for Supercapacitors." *Carbon*, 49(9), 2917–2925. doi:10.1016/j.carbon.2011.02.068.

Chu, Steven, and Arun Majumdar. 2012. "Opportunities and Challenges for a Sustainable Energy Future." *Nature*, 488(7411), 294–303. doi:10.1038/nature11475.

Conway, B. E. 1999a. "Electrochemical Capacitors Based on Pseudocapacitance." In *Electrochemical Supercapacitors*, 221–257. Boston, MA: Springer US. doi:10.1007/978-1-4757-3058-6_10.

Conway, B. E. 1999b. *Electrochemical Supercapacitors*. Boston, MA: Springer US. doi:10.1007/978-1-4757-3058-6.

Conway, B. E., and W. G. Pell. 2003. "Double-Layer and Pseudocapacitance Types of Electrochemical Capacitors and Their Applications to the Development of Hybrid Devices." *Journal of Solid State Electrochemistry*, 7(9), 637–644. doi:10.1007/s10008-003-0395-7.

Ding, Rui, Li Qi, Mingjun Jia, and Hongyu Wang. 2013. "Facile and Large-Scale Chemical Synthesis of Highly Porous Secondary Submicron/Micron-Sized $NiCo_2O_4$ Materials for High-Performance Aqueous Hybrid AC-$NiCo_2O_4$ Electrochemical Capacitors." *Electrochimica Acta*, 107(September), 494–502. doi:10.1016/j.electacta.2013.05.114.

Du, Wei, Yue-Ling Bai, Jiaqiang Xu, Hongbin Zhao, Lei Zhang, Xifei Li, and Jiujun Zhang. 2018. "Advanced Metal-Organic Frameworks (MOFs) and Their Derived Electrode Materials for Supercapacitors." *Journal of Power Sources*, 402(October), 281–295. doi:10.1016/j.jpowsour.2018.09.023.

Dubal, Deepak P., Pedro Gomez-Romero, Babasaheb R. Sankapal, and Rudolf Holze. 2015. "Nickel Cobaltite as an Emerging Material for Supercapacitors: An Overview." *Nano Energy*, 11(January), 377–399. doi:10.1016/j.nanoen.2014.11.013.

Fazal-ur-Rehman, Muhammad. 2018. "Methodological Trends in Preparation of Activated Carbon from Local Sources and Their Impacts on Production: A Review." *Chemistry International*, 4(2), 109–119. doi:10.5281/zenodo.1475350.

Feng, Xiao-Miao, Rui-Mei Li, Yan-Wen Ma, Run-Feng Chen, Nai-En Shi, Qu-Li Fan, and Wei Huang. 2011. "One-Step Electrochemical Synthesis of Graphene/Polyaniline Composite Film and Its Applications." *Advanced Functional Materials*, 21(15), 2989–2996. doi:10.1002/adfm.201100038.

Filleter, T., R. Bernal, S. Li, and H.D. Espinosa. 2011. "Ultrahigh Strength and Stiffness in Cross-Linked Hierarchical Carbon Nanotube Bundles." *Advanced Materials*, 23(25), 2855–2860. doi:10.1002/adma.201100547.

Gao, Xicheng, Weili Wang, Jianqiang Bi, Yafei Chen, Xuxia Hao, Xiaoning Sun, and Jingde Zhang. 2019. "Morphology-Controllable Preparation of $NiFe_2O_4$ as High Performance Electrode Material for Supercapacitor." *Electrochimica Acta*, 296(February), 181–189. doi:10.1016/j.electacta.2018.11.054.

González, Ander, Eider Goikolea, Jon Andoni Barrena, and Roman Mysyk. 2016. "Review on Supercapacitors: Technologies and Materials." *Renewable and Sustainable Energy Reviews*, 58(May), 1189–1206. doi:10.1016/j.rser.2015.12.249.

Graves, A. D., and D. Inman. 1965. "Adsorption and the Differential Capacitance of the Electrical Double-Layer at Platinum/Halide Metal Interfaces." *Nature*, 208(5009), 481–482. doi:10.1038/208481b0.

Guan, Bu, Akihiro Yuan, Le Kushima, Sa Yu, Ju Li, and Xiong Wen David Lou. 2017. "Coordination Polymers Derived General Synthesis of Multishelled Mixed Metal-Oxide Particles for Hybrid Supercapacitors." *Advanced Materials*, 29(17), 1605902. doi:10.1002/adma.201605902.

Habibzadeh, Mohamadhadi, Moeen Hassanalieragh, Akihiro Ishikawa, Tolga Soyata, and Gaurav Sharma. 2017. "Hybrid Solar-Wind Energy Harvesting for Embedded Applications: Supercapacitor-Based System Architectures and Design Tradeoffs." *IEEE Circuits and Systems Magazine*, 17(4), 29–63. doi:10.1109/MCAS.2017.2757081.

He, Yongmin, Wanjun Chen, Xiaodong Li, Zhenxing Zhang, Jiecai Fu, Changhui Zhao, and Erqing Xie. 2013. "Freestanding Three-Dimensional Graphene/MnO_2 Composite Networks As Ultralight and Flexible Supercapacitor Electrodes." *ACS Nano*, 7(1), 174–182. doi:10.1021/nn304833s.

Ho, M. Y., P. S. Khiew, D. Isa, T. K. Tan, W. S. Chiu, and C. H. Chia. 2014. "A Review Of Metal Oxide Composite Electrode Materials for Electrochemical Capacitors." *Nano*, 09(06), 1430002. doi:10.1142/S1793292014300023.

Hsu, Chun-Tsung, and Chi-Chang Hu. 2013. "Synthesis and Characterization of Mesoporous Spinel $NiCo_2O_4$ Using Surfactant-Assembled Dispersion for Asymmetric Supercapacitors." *Journal of Power Sources*, 242(November), 662–671. doi:10.1016/j.jpowsour.2013.05.130.

Hu, Liwen, Jiguo Tu, Shuqiang Jiao, Jungang Hou, Hongmin Zhu, and Derek J. Fray. 2012. "In Situ Electrochemical Polymerization of a Nanorod-PANI–Graphene Composite in a Reverse Micelle Electrolyte and Its Application in a Supercapacitor." *Physical Chemistry Chemical Physics*, 14(45), 15652. doi:10.1039/c2cp42192e.

Iro, Z. S. 2016. "A Brief Review on Electrode Materials for Supercapacitor." *International Journal of Electrochemical Science*, 11(December), 10628–10643. doi:10.20964/2016.12.50.

Jadhav, Harsharaj S., Sambhaji M. Pawar, Arvind H. Jadhav, Gaurav M. Thorat, and Jeong Gil Seo. 2016. "Hierarchical Mesoporous 3D Flower-like $CuCo_2O_4$/NF for High-Performance Electrochemical Energy Storage." *Scientific Reports*, 6(1), 31120. doi:10.1038/srep31120.

Kumbhar, Vijay S., Van Quang Nguyen, Yong Rok Lee, Chandrakant D. Lokhande, Do-Heyoung Kim, and Jae-Jin Shim. 2018. "Electrochemically Growth-Controlled Honeycomb-like $NiMoO_4$ Nanoporous Network on Nickel Foam and Its Applications in All-Solid-State Asymmetric Supercapacitors." *New Journal of Chemistry*, 42(18), 14805–14816. doi:10.1039/C8NJ02085J.

Lei, Yannick, Claire Fournier, Jean-Louis Pascal, and Frédéric Favier. 2008. "Mesoporous Carbon–Manganese Oxide Composite as Negative Electrode Material for Supercapacitors." *Microporous and Mesoporous Materials*, 110(1), 167–176. doi:10.1016/j.micromeso.2007.10.048.

Li, Jing, Huaqing Xie, Yang Li, Jie Liu, and Zhuxin Li. 2011. "Electrochemical Properties of Graphene Nanosheets/Polyaniline Nanofibers Composites as Electrode for Supercapacitors." *Journal of Power Sources*, 196(24), 10775–10781. doi:10.1016/j.jpowsour.2011.08.105.

Li, Lei, Jia Qin, Huiting Bi, Shili Gai, Fei He, Peng Gao, Yunlu Dai, Xitian Zhang, Dan Yang, and Piaoping Yang. 2017. "Ni(OH)$_2$ Nanosheets Grown on Porous Hybrid g-C$_3$N$_4$/RGO Network as High Performance Supercapacitor Electrode." *Scientific Reports*, 7(1), 43413. doi:10.1038/srep43413.

Li, Q. W., Y. Li, X. F. Zhang, S. B. Chikkannanavar, Y. H. Zhao, A. M. Dangelewicz, L. X. Zheng, et al. 2007. "Structure-Dependent Electrical Properties of Carbon Nanotube Fibers." *Advanced Materials*, 19(20), 3358–3363. doi:10.1002/adma.200602966.

Liu, Qiang, Junjie Yang, Xiaogang Luo, Yifei Miao, Yang Zhang, Wenting Xu, Lijun Yang, Yunxia Liang, Wei Weng, and Meifang Zhu. 2020. "Fabrication of a Fibrous MnO$_2$@MXene/CNT Electrode for High-Performance Flexible Supercapacitor." *Ceramics International*, 46(8), 11874–11881. doi:10.1016/j.ceramint.2020.01.222.

Lu, Junlin, Hao Ran, Jien Li, Jing Wan, Congcong Wang, Peiyuan Ji, Xue Wang, Guanlin Liu, and Chenguo Hu. 2020. "A Fast Composite-Hydroxide-Mediated Approach for Synthesis of 2D-LiCoO$_2$ for High Performance Asymmetric Supercapacitor." *Electrochimica Acta*, 331(mon), 135426. doi:10.1016/j.electacta.2019.135426.

Lu, Xianmao, and Younan Xia. 2006. "Buckling down for Flexible Electronics." *Nature Nanotechnology*, 1(3), 163–164. doi:10.1038/nnano.2006.157.

Lu, Xue-Feng, Dong-Jun Wu, Run-Zhi Li, Qi Li, Sheng-Hua Ye, Ye-Xiang Tong, and Gao-Ren Li. 2014. "Hierarchical NiCo$_2$O$_4$ Nanosheets@hollow Microrod Arrays for High-Performance Asymmetric Supercapacitors." *Journal of Materials Chemistry A*, 2(13), 4706–4713. doi:10.1039/C3TA14930G.

Luo, Jing, Sisi Jiang, Ren Liu, Yongjie Zhang, and Xiaoya Liu. 2013. "Synthesis of Water Dispersible Polyaniline/Poly(Styrenesulfonic Acid) Modified Graphene Composite and Its Electrochemical Properties." *Electrochimica Acta*, 96(April), 103–109. doi:10.1016/j.electacta.2013.02.072.

Luryi, Serge. 1988. "Quantum Capacitance Devices." *Applied Physics Letters*, 52(6), 501–503. doi:10.1063/1.99649.

Maldonado-Hódar, F. J., C. Moreno-Castilla, J. Rivera-Utrilla, Y. Hanzawa, and Y. Yamada. 2000. "Catalytic Graphitization of Carbon Aerogels by Transition Metals." *Langmuir*, 16(9), 4367–4373. doi:10.1021/la991080r.

Mao, Lu, Kai Zhang, Hardy Sze On Chan, and Jishan Wu. 2012. "Surfactant-Stabilized Graphene/Polyaniline Nanofiber Composites for High Performance Supercapacitor Electrode." *Journal of Materials Chemistry*, 22(1), 80–85. doi:10.1039/C1JM12869H.

Marcano, Daniela C., Dmitry V. Kosynkin, Jacob M. Berlin, Alexander Sinitskii, Zhengzong Sun, Alexander Slesarev, Lawrence B. Alemany, Wei Lu, and James M. Tour. 2010. "Improved Synthesis of Graphene Oxide." *ACS Nano*, 4(8), 4806–4814. doi:10.1021/nn1006368.

Miller, John R. 2013. "Reliability of Electrochemical Capacitors." In *Supercapacitors*, 473–507. Weinheim, Germany: Wiley-VCH Verlag GmbH & Co. KGaA. doi:10.1002/9783527646661.ch13.

Mini, P. A., Avinash Balakrishnan, Shanti V. Nair, and K. R. V. Subramanian. 2011. "Highly Super Capacitive Electrodes Made of Graphene/Poly(Pyrrole)." *Chemical Communications*, 47(20), 5753. doi:10.1039/c1cc00119a.

Mondal, Anjon Kumar, Dawei Su, Shuangqiang Chen, Katja Kretschmer, Xiuqiang Xie, Hyo-Jun Ahn, and Guoxiu Wang. 2015. "A Microwave Synthesis of Mesoporous NiCo$_2$O$_4$ Nanosheets as Electrode Materials for Lithium-Ion Batteries and Supercapacitors." *ChemPhysChem*, 16(1), 169–175. doi:10.1002/cphc.201402654.

Ngom, B. D., N. M. Ndiaye, N. F. Sylla, B. K. Mutuma, and N. Manyala. 2020. "Sustainable Development of Vanadium Pentoxide Carbon Composites Derived from Hibiscus

Sabdariffa Family for Application in Supercapacitors." *Sustainable Energy & Fuels*, 4(9), 4814–4830. doi: 10.1039/D0SE00779J.

Niu, Chunming, Enid K. Sichel, Robert Hoch, David Moy, and Howard Tennent. 1997. "High Power Electrochemical Capacitors Based on Carbon Nanotube Electrodes." *Applied Physics Letters*, 70(11), 1480–1482. doi: 10.1063/1.118568.

Pan, Yu, Hong Gao, Mingyi Zhang, Lu Li, Guangning Wang, and Xinyuan Shan. 2017. "Three-Dimensional Porous $ZnCo_2O_4$ Sheet Array Coated with Ni(OH)2 for High-Performance Asymmetric Supercapacitor." *Journal of Colloid and Interface Science*, 497(July), 50–56. doi: 10.1016/j.jcis.2017.02.053.

Pandolfo, A.G., and A.F. Hollenkamp. 2006. "Carbon Properties and Their Role in Supercapacitors." *Journal of Power Sources*, 157(1), 11–27. doi: 10.1016/j.jpowsour.2 006.02.065.

Park, S.-H., J.-Y. Kim, and K.-B. Kim. 2010. "Pseudocapacitive Properties of Nano-Structured Anhydrous Ruthenium Oxide Thin Film Prepared by Electrostatic Spray Deposition and Electrochemical Lithiation/Delithiation." *Fuel Cells*, 10(5), 865–872. doi: 10.1002/fuce.201000029.

Peigney, A., Ch. Laurent, E. Flahaut, R.R. Bacsa, and A. Rousset. 2001. "Specific Surface Area of Carbon Nanotubes and Bundles of Carbon Nanotubes." *Carbon*, 39(4), 507–514. doi: 10.1016/S0008-6223(00)00155-X.

Poonam, Kriti Sharma, Anmol Arora, and S.K. Tripathi. 2019. "Review of Supercapacitors: Materials and Devices." *Journal of Energy Storage*, 21(February), 801–825. doi: 10.1 016/j.est.2019.01.010.

Radhakrishnan, Logudurai, Julien Reboul, Shuhei Furukawa, Pavuluri Srinivasu, Susumu Kitagawa, and Yusuke Yamauchi. 2011. "Preparation of Microporous Carbon Fibers through Carbonization of Al-Based Porous Coordination Polymer (Al-PCP) with Furfuryl Alcohol." *Chemistry of Materials*, 23(5), 1225–1231. doi: 10.1021/cm102921y.

Rajkumar, Muniyandi, Chun-Tsung Hsu, Tzu-Ho Wu, Ming-Guan Chen, and Chi-Chang Hu. 2015. "Advanced Materials for Aqueous Supercapacitors in the Asymmetric Design." *Progress in Natural Science: Materials International*, 25(6), 527–544. doi: 10.1016/j.pnsc.2015.11.012.

Sari, Indah, Fitri Nur, and Jyh-Ming Ting. 2019. "High Performance Asymmetric Supercapacitor Having Novel 3D Networked Polypyrrole Nanotube/N-Doped Graphene Negative Electrode and Core-Shelled MoO_3/PPy Supported MoS_2 Positive Electrode." *Electrochimica Acta*, 320(October), 134533. doi: 10.1016/j.electacta.201 9.07.044.

Sharma, Meenu, and Anurag Gaur. 2020a. "Cu Doped Zinc Cobalt Oxide Based Solid-State Symmetric Supercapacitors: A Promising Key for High Energy Density." *The Journal of Physical Chemistry C*, 124(1), 9–16. doi: 10.1021/acs.jpcc.9b08170.

Sharma, Meenu, and Anurag Gaur. 2020b. "Designing of Carbon Nitride Supported $ZnCo_2O_4$ Hybrid Electrode for High-Performance Energy Storage Applications." *Scientific Reports*, 10(1), 2035. doi: 10.1038/s41598-020-58925-4.

Sharma, Meenu, Shashank Sundriyal, Amrish Panwar, and Anurag Gaur. 2018. "Enhanced Supercapacitive Performance of $Ni_{0.5}Mg_{0.5}Co_2O_4$ Flowers and Rods as an Electrode Material for High Energy Density Supercapacitors: Rod Morphology Holds the Key." *Journal of Alloys and Compounds*, 766(October), 859–867. doi: 10.1016/j.jallcom.201 8.07.019.

Sharma, R.K., A.C. Rastogi, and S.B. Desu. 2008. "Manganese Oxide Embedded Polypyrrole Nanocomposites for Electrochemical Supercapacitor." *Electrochimica Acta*, 53(26), 7690–7695. doi: 10.1016/j.electacta.2008.04.028.

Silva, Suse Botelho, Guilherme Lopes Batista, and Cristiane Krause Santin. 2019. "Chitosan for Sensors and Electrochemical Applications." In *Chitin and Chitosan*, 461–476. Wiley. doi: 10.1002/9781119450467.ch18.

Sivaraman, P., A. Thakur, R. K. Kushwaha, D. Ratna, and A. B. Samui. 2006. "Poly(3-Methyl Thiophene)-Activated Carbon Hybrid Supercapacitor Based on Gel Polymer Electrolyte." *Electrochemical and Solid-State Letters*, 9(9), A435. doi:10.1149/1.2213357.

Su, Li, Lijun Gao, Qinghua Du, Liyin Hou, Zhipeng Ma, Xiujuan Qin, and Guangjie Shao. 2018. "Construction of $NiCo_2O_4@MnO_2$ Nanosheet Arrays for High-Performance Supercapacitor: Highly Cross-Linked Porous Heterostructure and Worthy Electrochemical Double-Layer Capacitance Contribution." *Journal of Alloys and Compounds*, 749(June), 900–908. doi:10.1016/j.jallcom.2018.03.353.

Sugimoto, Wataru, Hideki Iwata, Katsunori Yokoshima, Yasushi Murakami, and Yoshio Takasu. 2005. "Proton and Electron Conductivity in Hydrous Ruthenium Oxides Evaluated by Electrochemical Impedance Spectroscopy: The Origin of Large Capacitance." *The Journal of Physical Chemistry B*, 109(15), 7330–7338. doi:10.1021/jp044252o.

Trasatti, Sergio, and Giovanni Buzzanca. 1971. "Ruthenium Dioxide: A New Interesting Electrode Material. Solid State Structure and Electrochemical Behaviour." *Journal of Electroanalytical Chemistry and Interfacial Electrochemistry*, 29(2), A1–A5. doi:10.1016/S0022-0728(71)80111-0.

Wang, Da-Yung, Chi-Lung Chang, and Wei-Yu Ho. 1999. "Microstructure Analysis of MoS2 Deposited on Diamond-like Carbon Films for Wear Improvement." *Surface and Coatings Technology*, 111(2–3), 123–127. doi:10.1016/S0257-8972(98)00712-9.

Wang, Guoping, Lei Zhang, and Jiujun Zhang. 2012. "A Review of Electrode Materials for Electrochemical Supercapacitors." *Chemical Society Reviews*, 41(2), 797–828. doi:10.1039/C1CS15060J.

Wang, Hualan, Qingli Hao, Xujie Yang, Lude Lu, and Xin Wang. 2009. "Graphene Oxide Doped Polyaniline for Supercapacitors." *Electrochemistry Communications*, 11(6), 1158–1161. doi:10.1016/j.elecom.2009.03.036.

Wang, H., Q. Hao, X. Yang, L. Lu, and X. Wang. 2010a. "A Nanostructured Graphene/Polyaniline Hybrid Material for Supercapacitors." *Nanoscale*, 2(10), 2164. doi:10.1039/c0nr00224k.

Wang, H., Q. Hao, X. Yang, L. Lu, and X. Wang. 2010b. "Effect of Graphene Oxide on the Properties of its Composite with Polyaniline." *ACS Applied Materials & Interfaces*, 2(3), 821–828. doi:10.1021/am900815k.

Wang, Huanlei, Chris M. B. Holt, Zhi Li, Xuehai Tan, Babak Shalchi Amirkhiz, Zhanwei Xu, Brian C. Olsen, Tyler Stephenson, and David Mitlin. 2012. "Graphene-Nickel Cobaltite Nanocomposite Asymmetrical Supercapacitor with Commercial Level Mass Loading." *Nano Research*, 5(9), 605–617. doi:10.1007/s12274-012-0246-x.

Wang, Junzhong, Kiran Kumar Manga, Qiaoliang Bao, and Kian Ping Loh. 2011. "High-Yield Synthesis of Few-Layer Graphene Flakes through Electrochemical Expansion of Graphite in Propylene Carbonate Electrolyte." *Journal of the American Chemical Society*, 133(23), 8888–8891. doi:10.1021/ja203725d.

Wang, Lu, Xiao Feng, Lantian Ren, Qiuhan Piao, Jieqiang Zhong, Yuanbo Wang, Haiwei Li, Yifa Chen, and Bo Wang. 2015. "Flexible Solid-State Supercapacitor Based on a Metal–Organic Framework Interwoven by Electrochemically-Deposited PANI." *Journal of the American Chemical Society*, 137(15), 4920–4923. doi:10.1021/jacs.5b01613.

Wang, Xu, Wan Shuang Liu, Xuehong Lu, and Pooi See Lee. 2012. "Dodecyl Sulfate-Induced Fast Faradic Process in Nickel Cobalt Oxide–Reduced Graphite Oxide Composite Material and Its Application for Asymmetric Supercapacitor Device." *Journal of Materials Chemistry*, 22(43), 23114. doi:10.1039/c2jm35307e.

Warren, Roseanne, Firas Sammoura, Fares Tounsi, Mohan Sanghadasa, and Liwei Lin. 2015. "Highly Active Ruthenium Oxide Coating via ALD and Electrochemical Activation in Supercapacitor Applications." *Journal of Materials Chemistry A*, 3(30), 15568–15575. doi:10.1039/C5TA03742E.

Wu, Lin, Li Sun, Xiaowei Li, Qiuyu Zhang, Haochen Si, Yuanxing Zhang, Ke Wang, and Yihe Zhang. 2020. "Mesoporous $ZnCo_2O_4$-CNT Microflowers as Bifunctional Material for Supercapacitive and Lithium Energy Storage." *Applied Surface Science*, 506(March), 144964. doi:10.1016/j.apsusc.2019.144964.

Wu, Wenling, Liuqing Yang, Suli Chen, Yanming Shao, Lingyun Jing, Guanghui Zhao, and Hua Wei. 2015. "Core–Shell Nanospherical Polypyrrole/Graphene Oxide Composites for High Performance Supercapacitors." *RSC Advances*, 5(111), 91645–91653. doi:1 0.1039/C5RA17036B.

Wu, Zhibin, Yirong Zhu, and Xiaobo Ji. 2014. "$NiCo_2O_4$-Based Materials for Electrochemical Supercapacitors." *Journal of Materials Chemistry A*, 2(36), 14759–14772. doi:10.1039/C4TA02390K.

Xu, Kaibing, Wenyao Li, Qian Liu, Bo Li, Xijian Liu, Lei An, Zhigang Chen, Rujia Zou, and Junqing Hu. 2014. "Hierarchical Mesoporous $NiCo_2O_4@MnO_2$ Core–Shell Nanowire Arrays on Nickel Foam for Aqueous Asymmetric Supercapacitors." *Journal of Materials Chemistry A*, 2(13), 4795. doi:10.1039/c3ta14647b.

Xue, Mianqi, Fengwang Li, Juan Zhu, Hang Song, Meining Zhang, and Tingbing Cao. 2012. "Structure-Based Enhanced Capacitance: In Situ Growth of Highly Ordered Polyaniline Nanorods on Reduced Graphene Oxide Patterns." *Advanced Functional Materials*, 22(6), 1284–1290. doi:10.1002/adfm.201101989.

Yang, L., W. Zheng, P. Zhang, J. Chen, W.B. Tian, Y.M. Zhang, and Z.M. Sun. 2018. "MXene/CNTs Films Prepared by Electrophoretic Deposition for Supercapacitor Electrodes." *Journal of Electroanalytical Chemistry*, 830–831(December), 1–6. doi:1 0.1016/j.jelechem.2018.10.024.

Yuan, Changzhou, Xiaogang Zhang, Linhao Su, Bo Gao, and Laifa Shen. 2009. "Facile Synthesis and Self-Assembly of Hierarchical Porous NiO Nano/Micro Spherical Superstructures for High Performance Supercapacitors." *Journal of Materials Chemistry*, 19(32), 5772. doi:10.1039/b902221j.

Zhang, Fang, Fei Xiao, Ze Hua Dong, and Wei Shi. 2013. "Synthesis of Polypyrrole Wrapped Graphene Hydrogels Composites as Supercapacitor Electrodes." *Electrochimica Acta*, 114(December), 125–132. doi:10.1016/j.electacta.2013.09.153.

Zhang, Honghao, Jun Wei, Yu Yan, Qianjin Guo, Liqiang Xie, Zhengchun Yang, Jie He, et al. 2020. "Facile and Scalable Fabrication of MnO_2 Nanocrystallines and Enhanced Electrochemical Performance of MnO_2/MoS_2 Inner Heterojunction Structure for Supercapacitor Application." *Journal of Power Sources*, 450(February), 227616. doi:1 0.1016/j.jpowsour.2019.227616.

Zhang, Li Li, Xin Zhao, Hengxing Ji, Meryl D. Stoller, Linfei Lai, Shanthi Murali, Stephen Mcdonnell, Brandon Cleveger, Robert M. Wallace, and Rodney S. Ruoff. 2012. "Nitrogen Doping of Graphene and Its Effect on Quantum Capacitance, and a New Insight on the Enhanced Capacitance of N-Doped Carbon." *Energy & Environmental Science*, 5(11), 9618. doi:10.1039/c2ee23442d.

Zhang, Sanliang, and Ning Pan. 2015. "Supercapacitors Performance Evaluation." *Advanced Energy Materials*, 5(6), 1401401. doi:10.1002/aenm.201401401.

Zhang, Yuehua, Ningke Hao, Xuejiao Lin, and Shuangxi Nie. 2020. "Emerging Challenges in the Thermal Management of Cellulose Nanofibril-Based Supercapacitors, Lithium-Ion Batteries and Solar Cells: A Review." *Carbohydrate Polymers*, 234(April), 115888. doi:10.1016/j.carbpol.2020.115888.

Zheng, Shasha, Xinran Li, Bingyi Yan, Qin Hu, Yuxia Xu, Xiao Xiao, Huaiguo Xue, and Huan Pang. 2017. "Transition-Metal (Fe, Co, Ni) Based Metal-Organic Frameworks for Electrochemical Energy Storage." *Advanced Energy Materials*, 7(18), 1602733. doi:10.1002/aenm.201602733.

5 Electrolytes for Li-Ion Batteries and Supercapacitors

Anil Arya[1,2], Lokesh Pandey[3], Anurag Gaur[2], Vijay Kumar[4], and A.L. Sharma[1]
[1]Department of Physics, Central University of Punjab, Bathinda 151401, India
[2]Department of Physics, National Institute of Technology, Kurukshetra 136119, Haryana, India
[3]Department of Physics, Uttrakhand Technical University, Dehradun, Uttrakhand, India
[4]Department of Physics, Institute of Integrated and Honors Studies (IIHS), Kurukshetra University, Kurukshetra 136119, India

CONTENTS

DOI: 10.1201/9781003141761-5

5.1   INTRODUCTION

Energy plays an important role in daily human life and is the basic need in the present scenario. Nowadays, the development of energy sources is linked with the development of human civilization. However, the increasing demand for energy and diminution of fossil fuels have worried the scientific/industry community to look for alternative energy resources. It is a need for time to switch toward renewable sources of energy owing to the challenging issues with nonrenewable sources, e.g. increased global warming, air/soil/water pollution, and limited stock of existing natural material. Therefore, the top priority of the researchers is now to develop sustainable and environmentally friendly sources of energy with an overall collective approach to address the future energy demand in portable electronics, military/space operations, household supply, electric vehicles, and power grids. The two most important devices are battery and supercapacitor (SC) that bear the potential to bridge the storage and production of renewable energy to fulfilling the increased energy demand. Thus, the high energy and power density devices are on the radar of the research community [Yan et al. 2014, Dubal et al. 2015]. In battery and supercapacitor, important constituents are electrodes (cathode and anode), separator, and electrolyte. An electrolyte is an essential constituent in the energy storage device and fully controls the charging as well as discharging of the cell [Kim et al. 2017]. It may be in the form of the liquid, quasi-solid, solid, and its key role is to act as a medium for ion migration (ionically conducting and electronically insulating). It prevents the electron flow within, as electrons flow through an external load. In the cell assembly, the electrolyte is sandwiched between the two electrodes (cathode and anode). Thus, the electrolyte plays a key role in deciding the cell performance as well as the safety of the device [Arya and Sharma 2017a,Tarascon et al. 2017].

5.2   IMPORTANT FEATURES OF EFFICIENT ELECTROLYTE

The electrolyte is a crucial component for the battery and supercapacitor and releases ions for conduction. It also acts as a carpet for the ions and hinders the electron migration within. The electrolyte needs to fulfill some specific requirements for its utility in the fabrication of the device. In existing technology using liquid electrolytes, separators are also used for preventing contact with electrodes. However, this component can be eliminated by using a solid electrolyte (separator cum electrolyte) that plays a dual role. The important features that decide the electrolyte suitability are its conductivity (ionic and electronic), stability (thermal, mechanical, and chemical), cost, thickness, and weight (Figure 5.1). Table 5.1 summarizes some key requirements for separators for Li-ion batteries.

5.2.1   IONIC CONDUCTIVITY

Ionic conductivity is an important parameter for electrolytes and decides the overall cell performance. High ionic conductivity and negligible electronic conductivity are the keys to achieve high device performance. In general, for aqueous liquid electrolytes, the ionic conductivity is generally high owing to better dissociation of salt and low viscosity. Whereas the conductivity of the

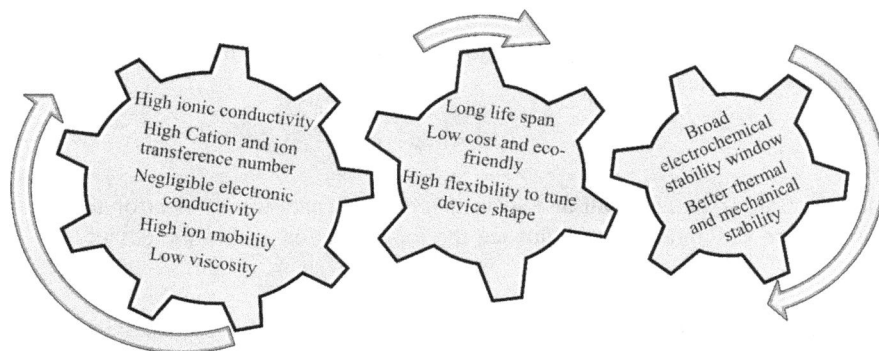

FIGURE 5.1 Key characteristics of electrolyte for battery and supercapacitor.

nonaqueous liquid electrolyte is lower than that of aqueous electrolyte owing to the higher viscosity of solvents. The ionic liquid-based polymer electrolyte has conductivity in the desirable range for device applications, followed by gel polymer electrolytes. The conductivity is inferior for the solid polymer electrolytes, but it can be improved by the incorporation of nanofillers (active and passive), nanoclay intercalation, nanowire addition, etc. Therefore, conductivity is an important parameter that needs to be enhanced for electrolytes and should be of the order of 10^{-3} S/cm at ambient/sub-ambient temperature [Arya and Sharma 2017a, Arya and Sharma 2017b]. The ion dynamics in polymer electrolytes are via the hopping of cation through coordinating sites provided by the electron-rich group in polymer chains. The flexibility of the polymer chain promotes faster ion dynamics and hence the ionic conductivity [Joost et al. 2015].

TABLE 5.1

General Requirements for Separators Used in Lithium-Ion Batteries

Parameters	Requirements
Chemical and electrochemical stabilities	Stable for an extended period
Wettability	Wet out quickly and completely
Mechanical property	>1000 kg/cm (98.06 MPa)
Thickness	20–25 μm
Pore size	<1 μm
Porosity	40–60%
Permeability (Gurley)	<0.025 sec/μm
Dimensional stability	No curl up and lay flat
Thermal stability	<5% shrinkage after 60 minutes at 90 °C
Shutdown	Effectively shut down the battery at elevated temperatures

Source: Reproduced with permission from Lee et al. (2014), Royal Society of Chemistry (Great Britain).

The conductivity of the electrolyte (σ) is directly linked to charge (z), number of free ions (n), and ion mobility (μ). The conductivity of polymer electrolyte is expressed by the equation

$$\sigma = ne\mu \qquad (5.1)$$

The selection of suitable salt and solvent for the preparation of electrolytes is necessary. The salt and solvent influence the ion migration within the device.

5.2.1.1 Characteristics of Salt

The cation and anion radii and lattice energy influence the conductivity of the electrolyte. Therefore, during the selection of electrolytes, it becomes important to examine the nature of the salt. The salt must have some characteristic features such as low lattice energy, high ionic conductivity, high ion mobility, broad electrochemical voltage stability window, and large anion size. For better comparison and selection of salt, Table 5.2 summarizes the crucial properties and provides a detailed comparison of various salts.

5.2.1.2 Characteristics of Solvent

The solvent also influences the conductivity of the electrolyte. The dielectric constant and viscosity are two factors that decide the ion dynamics as well as salt dissociation. The best electrolyte should have a high dielectric constant, low viscosity, low melting point, and the ability to form a complex.

The cation migration in polymer-based electrolytes is associated with the flexibility of polymer chains and their glass transition temperature (T_g). Temperature alters the flexibility of polymer chains, and with an increase in temperature increase in conductivity is observed. This increase is due to the thermal activation of charge carriers and the decrease of the activation barrier. These thermally activated charge carriers easily overcome the barrier height and participate in conduction. The

TABLE 5.2

Comparison between Salts for Lithium Batteries

Properties	From best to worst →					
Ion mobility	$LiBF_4$	$LiClO_4$	$LiPF_6$	$LiAsF_6$	LiTf	LiTFSI
Ion pair dissociation	LiTFSI	$LiAsF_6$	$LiPF_6$	$LiClO_4$	$LiBF_4$	LiTf
Solubility	LiTFSI	$LiPF_6$	$LiAsF_6$	$LiBF_4$	LiTf	-
Thermal stability	LiTFSI	LiTf	$LiAsF_6$	$LiBF_4$	$LiPF_6$	-
Chemical inertness	LiTf	LiTFSI	$LiAsF_6$	$LiBF_4$	$LiPF_6$	-
SEI formation	$LiPF_6$	$LiAsF_6$	LiTFSI	$LiBF_4$	-	-
Al corrosion	$LiAsF_6$	$LiPF_6$	$LiBF_4$	$LiClO_4$	LiTf	LiTFSI

Source: Tasaki et al. (2003) and Mauger et al. (2018).

variation of ionic conductivity with temperature is explained by two commonly referred mechanisms: (i) Arrhenius behavior (for low temperature) and (ii) Vogel–Tammann–Fulcher (VTF) behavior (for high temperature). The activation energy (E_a) is estimated, and low activation energy, as well as low glass transition temperature (T_g), suggests optimum ionic conductivity required for any electrolyte.

5.2.1.3   Arrhenius Behavior

With an increase in temperature, polymer flexibility increases, and the crystalline phase suppresses that support the electrolyte with great feasibility. The availability of amorphous content and enhanced chain flexibility promotes faster ion migration. The Arrhenius behavior is expressed as

$$\sigma = \sigma_o \exp\left(-\frac{E_a}{kT}\right) \tag{5.2}$$

Here, σ_o is the pre-exponential factor, k is the Boltzmann constant, and E_a is the activation energy.

5.2.1.4   Vogel–Tammann–Fulcher (VTF) Behavior

In this mechanism, the segmental motion of the polymer chain favors the ion dynamics inside the polymer matrix by providing free volume to cation. It depicts long-range ion dynamics via hopping. The VTF behavior is expressed as

$$\sigma = AT^{-1/2} \exp\left(-\frac{B}{T - T_o}\right) \tag{5.3}$$

Here, σ is the ionic conductivity, A is the pre-exponential factor related to the conductivity and number of free charge carriers, B is the pseudoactivation energy for the conductivity, and T_o is the temperature (close to the T_g).

5.2.2   Cation Transference Number

When salt is added to the polymer–solvent system, then it gets dissociated due to interaction with polymer chains and solvent. The cations and anions are released. The more the number of free cations, the more will be the ionic conductivity. Thus, proper salt dissociation is required for more free cations. This is investigated in terms of cation transference number (t_+) and for the ideal case, its value is unity. Both cation and anion contribute to conductivity, not cation plays a dominant role. The cation contribution is evaluated using the following equation (Bruce–Vincent equation):

$$t_{Li^+} = \frac{I_s(V - I_i R_i)}{I_i(V - I_s R_s)} \tag{5.4}$$

Here, V is the applied voltage, I_i and I_s are the initial and steady-state currents, and R_i and R_s are the interfacial resistance before and after polarization. The ion transference number (t_{ion}) is obtained via the equation: $t_{ion} = \frac{(I_t - I_e)}{I_t} \times 100$; [$I_t$ and I_e are the total current and the residual current, respectively, and are related as $i_t = i_{ion} + i_{elec}$].

5.2.3 ELECTROCHEMICAL STABILITY WINDOW

The energy density and capacity are linked with each other as $E = 0.5\ CV^2$. Voltage plays an important role, and it indicates the operating voltage range for the device. The voltage stability window is evaluated with the linear sweep voltammetry (LSV) technique. For application purposes, the voltage window of the electrolyte must be >4 V.

5.2.4 ION TRANSPORT PARAMETERS

Along with conductivity and voltage stability, ion transport parameters also influence ion dynamics. The study of these provides detailed ion dynamics for the electrolyte. Diffusion constant (D), mobility (μ), and charge carrier concentration (n) are key parameters and can be calculated via different methods (Table 5.3) [Bandara et al. 2011, Arof et al. 2014].

TABLE 5.3

Approaches to Obtain the Number Density (n), Mobility (μ), and Diffusion Coefficient (D)

Method units	Bandara and Mellander (B-M) approach	Impedance spectroscopy approach	FTIR method
D (cm^{2}s^{-1})	$D = \frac{d^2}{\tau_2\delta^2}$	$D = \frac{(k_2\varepsilon_r\varepsilon_0 Ad)^2}{\tau_2}$	$D = \frac{\mu k_B T}{e}$
μ (cm^{2}V^{-1}s^{-1})	$\mu = \frac{ed^2}{kT\tau_2\delta^2}$	$\mu = \frac{e(k_2\varepsilon_r\varepsilon_0 Ad)^2}{k_B T\tau_2}$	$\mu = \frac{\sigma}{ne}$
N (cm^{-3})	$n = \frac{\sigma kT\tau_2\delta^2}{e^2 d^2}$	$n = \frac{\sigma k_B T\tau_2}{(ek_2\varepsilon_r\varepsilon_0 Ad)^2}$	$n = \frac{M \times N_A}{V_{Total}} \times$ freeionarea(%)
Parameters	τ_2 is a time constant corresponding to the maximum dissipative loss curve, $\delta = d/\lambda$, λ is the thickness of the electrical double layer, and d is half-thickness of the polymer electrolyte.	k_2 and k_1 are obtained from the trial and error method on the Nyquist plot. The value of τ_2 was taken at the frequency corresponding to a minimum in the imaginary parts of the impedance, Z_i, i.e. at $Z_i \to 0$, k_B is the Boltzmann constant (1.38 × 10^{-23} J K^{-1}), and T is the absolute temperature.	M is the number of moles of salt used in each electrolyte, N_A is Avogadro's number (6.02 × 10^{23} mol^{-1}), V_{Total} is the total volume of the solid polymer electrolyte, and σ is dc conductivity, e is the electric charge (1.602 × 10^{-19} C), k_B is the Boltzmann constant (1.38 × 10^{-23} J K^{-1}), and T is the absolute temperature.

5.2.5 Thermal Stability

The thermal stability of an electrolyte is a crucial parameter as it decides the safe operation of any device. As the electrolyte is sandwiched between electrodes and prevents short-circuiting, therefore, it needs to safely stand working for a broad temperature range (−20 to 80 °C). As existing devices are using organic liquid as an electrolyte, therefore, during rapid cell operation it decomposes and releases heat that escalates temperature as well as pressure inside the cell. The thermal stability of solid-state electrolytes is better as compared to conventional liquid electrolytes. Further, both conductivity and thermal stability can be improved by the proper selection of nanofiller. Wu et al. (2020) used a flame retardant rod-type nanofiller [$Zn_2(OH)BO_3$] and were able to enhance the thermal stability close to 420 °C and ionic conductivity (2.78×10^{-5} S/cm at 30 °C) for solid polymer electrolytes. The voltage stability window was 4.51 V. Also, the thermal runaway (TR) in one cell damages it, and TR may propagate in the whole battery stack and also catch fire. That is why solid electrolytes of high ionic conductivity are being developed to replace the liquid electrolyte, and it will also open opportunities to develop an all-solid-state li-ion battery (ASSLIB) or supercapacitor. Figure 5.2a shows the thermal runaway process based on the degree of internal short or heat generation speed [Feng et al. 2019]. TR may be the result of mechanical abuse (separator damage may trigger short-circuit), electrical abuse (short-circuit releases heat and internal pressure may rise), and thermal abuse (decomposition of electrode and electrolyte and high temperature inside the cell). An internal short-circuit in the cell connects the cathode and anode that results in rapid heat generation [Wen et al. 2012]. Figure 5.2b shows the failure mechanism due to an internal short-circuit.

5.3 CLASSIFICATION OF DIFFERENT ELECTROLYTES

Two important energy storage/conversion devices are battery and supercapacitor. The electrolyte is an integral part of any device and plays an important role during the charging/discharging operation. Different types of electrolytes have been used based on their conductivity, stability, and safety properties. During the selection of electrolytes, an optimum balance of three properties (conductivity, thermal/mechanical stability, and safety) needs to be achieved. For better understanding, the electrolytes are classified into various types and advantages are also highlighted (Figure 5.3).

5.3.1 Aqueous/Nonaqueous Liquid

An organic electrolyte system comprises an organic solvent and a supporting electrolyte. The important features of organic electrolytes are high dielectric constant, low volatility, and better electrochemical stability. The important point to have an organic electrolyte for achieving optimum performance is a selection of organic solvents and their compatibility with electrode materials. For high performance of energy devices, water as a solvent is not preferred. Important parameters of different electrolyte solvents are summarized in Table 5.4.

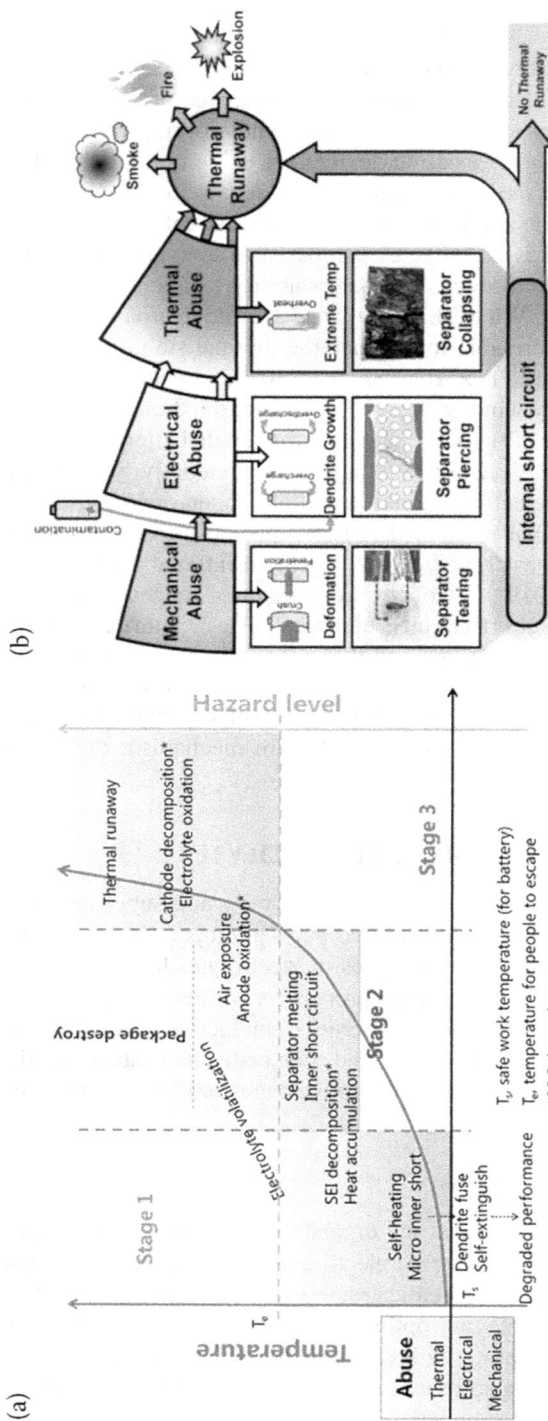

FIGURE 5.2 (a) Three stages in battery TR, T_s, safe work temperature (for battery); T_e, temperature for people to escape; *SOC dependent. [Reproduced with permission from Wu et al. 2019]. (b) Internal short-circuit: the most common feature of TR [Reproduced with permission from Feng et al. 2018].

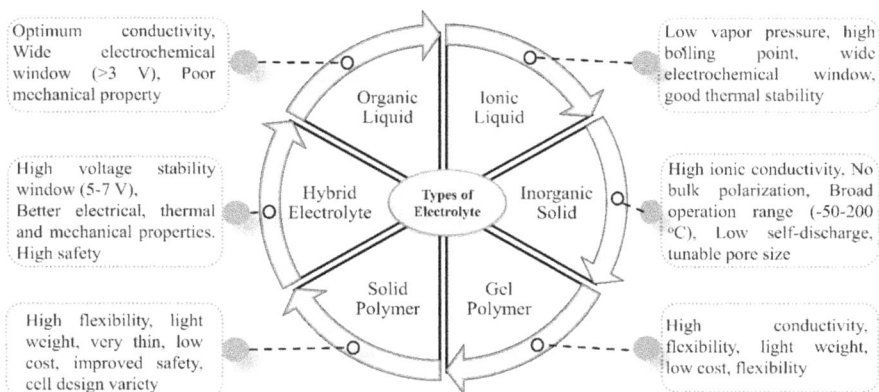

Optimum conductivity, Wide electrochemical window (>3 V), Poor mechanical property

Low vapor pressure, high boiling point, wide electrochemical window, good thermal stability

High voltage stability window (5-7 V), Better electrical, thermal and mechanical properties. High safety

High ionic conductivity. No bulk polarization, Broad operation range (-50-200 °C), Low self-discharge, tunable pore size

High flexibility, light weight, very thin, low cost, improved safety, cell design variety

High conductivity, flexibility, light weight, low cost, flexibility

Organic Liquid — Ionic Liquid — Hybrid Electrolyte — Types of Electrolyte — Inorganic Solid — Solid Polymer — Gel Polymer

FIGURE 5.3 Classification of different electrolytes.

The electrochemical performance of the supercapacitor cell based on Bi_2WO_6 nanoparticles as an electrode with different electrolytes (1 M NaOH, 1 M LiOH, 1 M Na_2SO_4, 1 M KOH, and 6 M KOH) was examined by Nithya et al. (2013). The KOH-based cell exhibits the highest specific capacitance of 304 F/g (at 3 mA/cm²). This enhancement is due to superior ionic mobility and low equivalent series resistance. The high value of current response for KOH (6 M KOH > 1 M KOH > 1 M NaOH > 1 M Na_2SO_4) is attributed to the difference in the hydration sphere radius of different ions (K^+, 3.31 Å; Na^+, 3.58 Å; Li^+, 3.82 Å). Also, the conductivity of the K^+ (73 cm²/Ω) is superior to Li^+ (38 cm²/Ω) and Na^+ (50 cm²/Ω). Wang et al. (2017a) examined the effect of organic electrolyte and ionic liquid electrolyte on carbon electrodes for supercapacitor. The SC cell with organic electrolyte demonstrates the specific capacitances of 146 F/g and IL electrolyte exhibits 224 F/g (at 0.1 A/g). The energy densities were 26 Wh/kg (organic electrolyte) and 92 W h/kg (IL electrolyte). Table 5.5 compares the electrochemical performance for various systems.

The colloidal nanocrystal clusters (CNCs) of $MnFe_2O_4$ were used as an electrode and tested in different aqueous electrolytes (KOH, NaOH, LiOH, and Na_2SO_4) by Wang et al. 2014. The specific capacitance was highest for the KOH electrolyte-based cell followed by NaOH, LiOH, and $LiNO_3$ (Figure 5.4i,ii). The specific capacitances increase with the decrease in hydrated ionic radii. The smaller cation radius suggests stronger polarization that increases the radius of hydrated ions (Figure 5.4iii). This hinders the electrolyte ion migration and hence reduction in capacitance.

5.3.2 IONIC LIQUID

IL-based polymer electrolytes are alternatives to traditional electrolytes. ILs are molten salts and are liquid below 100 °C due to weakly coordinated ions. The broad electrochemical potential window, high conductivity, ion mobility, nontoxic nature and non-flammability, nonvolatility, excellent thermal stability, and low vapor pressure are some

TABLE 5.4
Common Electrolyte Solvents in Energy Storage Devices

Name	Structure	M_w	T_m (°C)	T_b (°C)	η (cP, °C)	ε (25 °C)	Dipole moment (Debye)	T_f (°C)	d (g mL^{-1}, °C)
Ethylene carbonate (EC)		88	36.4	248	1.90 (40 °C)	89.78	4.61	160	1.321
Propylene carbonate (PC)		102	−48.8	242	2.53	64.92	4.81	132	1.200
Dimethyl carbonate (DMC)		90	4.6	91	0.59 (20 °C)	3.107	0.76	18	1.063
Diethyl carbonate (DEC)		118	−74.3	126	0.75	2.805	0.96	31	0.969
Ethyl methyl carbonate (EMC)		104	−53	110	0.65	2.958	0.89	27	1.006
γ-Butyrolactone (GBL)		86	−43.5	204	1.73	39	4.23	97	1.199

Source: Reproduced with permission from Xu (2004), American Chemical Society.

TABLE 5.5

Electrochemical Performance for Different Electrolyte-based Supercapacitors

Electrolyte	Electrode	Voltage (V)	Specific capacitance (F/g)	Cyclic stability	References
1 M Na_2SO_4	CNF//CNF	0.8	69.3 at 50 mV/s	2000 cycles	Oyedotun et al. 2019
1.5 M Na_2SO_4	AC//AC	1.0	93.1 at 0.005 A/g	-	Tey et al. 2016
6 M KOH	S-CB//S-CB	1.0	120 at 1 A/g	92.6% after 10000 cycles	Ma et al. 2019
1M $TEABF_4$/ACN	Highly porous interconnected carbon nanosheets	2.7	120–150 at 1 mV/s	-	Sevilla and Fuertes 2014
1 M $TEABF_4$/PC	Graphene–CNT composites	3	110 at 1 A/g	-	Jung et al. 2013
0.5 M K_2SO_4	AC//AC	1.7	42.8 at 0.1 A/g	96% after 1000 cycles	Chavhan and Ganguly 2017
3 M H_2SO_4	CNT-MC//CNt-MC	0.8	237 at 1 A/g	92% after 20000 cycles	Yao et al. 2015
1 M $NaPF_6$/(EC–DMC–PC–EA1: 1: 1 : 0.5)	Microporous carbide-derived carbon	3.4	120 at 1 mV/s	-	Väli et al. 2014
1 M H_2SO_4	ANS–rGO	2.0	375 at 1.3 A/g	-	Jana et al. 2014

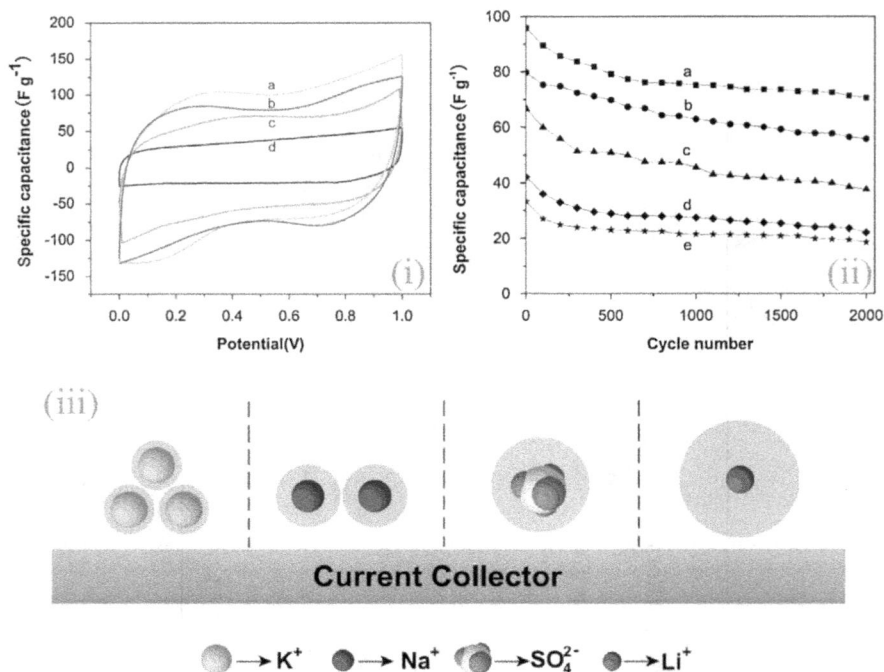

FIGURE 5.4 (i) CV curves of the SC cell with the electrolytes (2 M) of (a) KOH, (b) NaOH, (c) LiOH, and (d) Na$_2$SO$_4$ at the scan rates of 5 mV/s. (ii) Cycle stability for various electrolyte: (a) 6 M KOH, (b) 2 M KOH, (c) 2 M LiOH, (d) 0.5 M KOH, (e) 2 M Na$_2$SO$_4$. (iii) The ion of hydrated ionic radius on the MnFe$_2$O$_4$-based supercapacitor [Reproduced with permission from Wang et al. 2014].

key features of any IL. IL also acts as an additive in polymer electrolytes and favors salt dissociation. One key advantage is that IL has a tunable structure that helps in fabricating the device with a broad temperature range. Some examples are EMITf, EMIM-TY, BMPyTFSI PYR$_{13}$FSI EMIMTFSI, and BMITFSI. Composite materials (blending organic and inorganic) can be prepared easily by using IL owing to the simultaneous presence of electrostatic charges, aromatic groups, and alkyl segments in the ion structures. This facilitates the enhanced interfacial properties owing to tunable nanostructures that are critical for any device [Yang et al. 2018]. A polymer electrolyte with PEO+ LiDFOB polymer matrix and 1-ethyl-3-methylimidazolium bis(trifluoromethylsulfonyl)imide (EMImTFSI) ionic liquid was reported [Polu and Rhee 2017]. The highest ionic conductivity was 1.85×10^{-4} S/cm. The initial specific capacity for the fabricated cell was 155 mAh/g. The PEO–LiTFSIpolymer matrix with different N-alkyl-N-methylpyrrolidinium bis(trifluoromethanesulfonyl)imide (PYR1ATFSI) ionic liquids was reported by Kim et al. (2007). The ionic conductivity value was about 10^{-4} S/cm, and the battery test of cell Li/GPE/LiFePO$_4$ shows a capacity of 125 mAh/g (at 30 °C) and 100 mAh/g (at 30 °C). Balo et al. (2017) reported the preparation of a GPE based on PEO-LiTFSI with EMIMTFSI ionic liquid. The highest ionic conductivity was

2.08 × 10^{-4} S/cm at 12.5 wt % ILs. Optimum electrolyte shows the highest ion transference number ($t_{ion} > 0.99$) and cation transference number ($t_+ = 0.39$). The electrochemical stability window was around 4.6 V. The fabricated cell shows a discharge capacity of 56 mAh/g (at C/10 for the first cycle) and 120 mAh/g in the tenth cycle with an efficiency (η) of 98 % (after 100 cycles). The ionic conductivity reported for polymer electrolyte P(EO)$_{20}$LiTFSI+1-butyl-4-methyl pyridinium bis(trifluoromethanesulfonyl) imide (BMPyTFSI) at 40 °C was 6.9 × 10^{-4} S/cm [Cheng et al. 2007]. The electrochemical stability window after IL addition was 5.3 V and is higher than the polymer salt system (4.8 V).

The addition of ionic liquid N-methyl-N-propyl piperidinium bis (trifluoromethanesulfonyl) imide (PP1.3TFSI), in the PEO-LiTFSI-based polymer matrix, demonstrates the ionic conductivity of about ~2.06 × 10^{-4} S/cm [Yongxin et al. 2012]. The voltage stability window was 4.5–4.7 V (versus Li/Li$^+$) and t_+ was 0.339. The cell shows capacity of about 120 mAh/g and coulomb efficiency was greater than 99% for 20 cycles. Another work with the addition of 1-butyl-3-methylimidazolium methylsulfate (BMIM-MS) ionic liquid in PEO-NaMS (sodium methylsulfate) was reported by Singh et al. (2016). The ionic conductivity value was 1.05 × 10^{-4} S/cm for 60 wt. % of IL loading ($t_+= 0.46$). The electrochemical stability window was in the range of 4–5 V.

A quaternary, PEO-LiTFSI-based polymer electrolyte was synthesized with N-methyl-N-propylpyrrolidinium bis(fluorosulfonyl)imide (PYR$_{13}$FSI) ionic liquid [Simonetti et al. 2017]. The conductivity value was 3.4 × 10^{-4} (−20 °C), 2.43 × 10^{-3} S/cm (20 °C), and 9.1 × 10^{-3} S/cm (60 °C). This increase in conductivity with IL content was owing to the formation of a 3D network of highly conductive IL pathways. The voltage stability window was 4.5 V. Addition of EMImTFSI ionic liquid was investigated on the (PEO)$_8$LiTFSI-10% NC-based polymer matrix. The highest ionic conductivity obtained was the order of 10^{-2} S/cm (at 343 K) for 10 wt % IL content (voltage stability window = 3.97 V) [Karuppasamy et al. 2016].

5.3.3 Inorganic Solid

Inorganic solid electrolytes (Garnet, Perovskite, LISCION, NASCION, etc.) are alternatives to liquid electrolytes as they allow fast Li-ion conduction. Other important features are (i) high ionic conductivity (~10^{-4}–10^{-3} S/cm, (ii) broad electrochemical voltage stability window (0–6 V), (ii) enhanced safety, and (iv) better thermal, mechanical stability, and chemical compatibility. Also, inorganic solid electrolyte-based battery and supercapacitor demonstrate high specific capacitance (or specific capacity), better cyclic stability, and high energy/power density. Due to the solid nature of electrolytes, the only challenge is to improve the interfacial contact for high device performance. Various schemes have been implemented to improve the interfacial contact with electrodes [Choi et al. 2007, Famprikis et al. 2019, Bachman et al. 2016, Cao et al. 2014]. Li *et al.,* 2019 fabricated the supercapacitor cell using carbon nanofoam from biowaste as an electrode and high-voltage inorganic gel electrolyte. The fabricated supercapacitor cell has a specific capacitance of 177 F/g. Further, the CMC-Na/Na$_2$SO$_4$ gel electrolyte was used to fabricate the A-CS650//A-CS650 symmetric device. All CV curves are quasi-rectangular at 10 mV/s (Figure 5.5a). However, at a

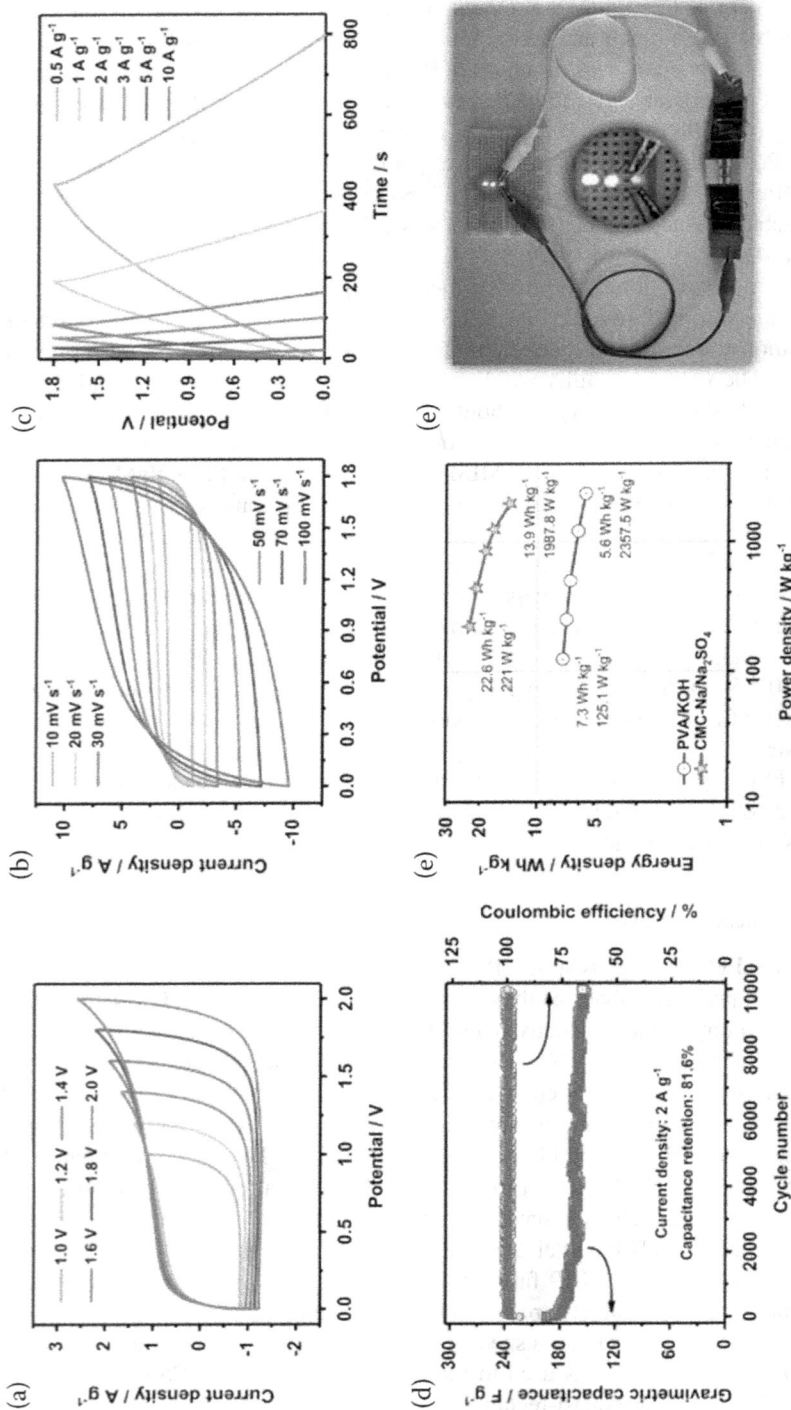

FIGURE 5.5 Electrochemical performance of the QSSC device with CMC-Na/Na$_2$SO$_4$ gel electrolyte: (a) CV curves at a scan rate of 10 mV/s with various potential ranges, (b) CV curves at various scan rates, (c) GCD curves at various current densities, (d) cyclability and Coulombic efficiency at 2 A/g and (e) Ragone plots; (f) three LED bulbs lighted by two devices connected in series [Reproduced with permission from Li et al., 2019b, Elsevier].

large window (~2.0 V), an electrolyte decomposition signature appears. Figure 5.5b shows the CV curve at different scan rates, and Figure 5.5c shows the GCD (Galvanostatic Charge-Discharge) curve at different current densities. The fabricated device demonstrates 81.6 % capacity retention and 100% Coulombic efficiency after 10000 cycles (Figure 5.5d). Ragone plots for two different electrolytes are shown in Figure 5.5e. The energy density for CMC-Na/Na$_2$SO$_4$ electrolyte is 22.6 Wh/kg and power density is 221 W/kg. Figure 5.5f demonstrates that the three light-emitting diodes (LEDs) bulbs glow.

The inorganic electrolyte has boosted the operation range of the battery attributes to the high ionic conductivity and better stability. LISICON, Garnet, and NASICON type electrolytes emerged as alternatives to liquid and gel electrolytes. Kamaya et al. (2011) prepared a 3D framework structure of Li$_{10}$GeP$_2$S$_{12}$, and it demonstrates high conductivity of 12 m S/cm at 27 °C. Another attractive inorganic electrolyte category is garnet-type, and Li$_7$La$_3$Zr$_2$O$_{12}$ (LLZO) is being investigated. The lithium ions migrate within the garnet lattice framework with a 3D conduction mechanism [Murugan et al. 2007, Dumon et al. 2013]. Allen et al. (2012) prepared Li$_{6.75}$La$_3$Zr$_{1.75}$Ta$_{0.25}$O$_{12}$ cubic garnet, and it exhibits the highest conductivity of about 8.7 × 10^{-4} S/cm (at 25 °C). Another report [Jin et al. 2013] presents the use of LLZO as an electrolyte for the fabrication of all-solid-state batteries (Cu$_{0.1}$V$_2$O$_5$/LLZO/Li), and initial discharging capacity was 93 mAh/g at 10 µA/cm^2 (at 50 °C). The full cell demonstrates a discharge capacity of about 129 mAh/g (for the first cycle) and 127 mAh/g (for the 100th cycle). Another report by Bron et al. (2013) prepared the Li$_{10}$SnP$_2$S$_{12}$ by replacing Ge with Sn. The highest conductivity value was 4 mS/cm (at RT). Then, an all-solid-state battery (ASSBs) was fabricated using Li$_{10}$GeP$_2$S$_{12}$ (cathode: LiCoO$_2$; anode: in metal) as an electrolyte, and the cell shows discharge capacity of about 120 mAh/g (Coulombic efficiency = 100%).

5.3.4 GEL POLYMER

The development of gel polymer electrolytes (GPE) has opened new doors of opportunities and has the potential to replace traditional electrolytes (organic electrolytes). Enhanced stability (thermal and mechanical) properties for GPE strengthens its candidature for fabricating a safe battery. A cross-linked GPE porous fabric membrane (XSAE) exhibits ionic conductivity of 0.67 mS/cm (at 30 °C) [Tsao 2015]. This enhancement in conductivity is attributed to the largest electrolyte uptake (85.1%). The t_+ was 0.58 and the voltage stability window was 4.5 V. The fabricated device has a discharge capacity of 154 mAh/g (at 0.1 °C) and 145 mAh/g (at 1 °C). This energy density was better than the XAE system (PDMS free). PVdF-co-HFP (electrospinning and nonwoven) based polymer electrolytes have gained attention due to unique features, such as (i) flexibility, (ii) broad voltage stability window, and (iii) nonflammable nature, However, the leakage of electrolyte is an issue that restricts the use for application [Zhu et al. 2013, Pitawala et al. 2014]. Therefore, to eliminate this issue, an oligomeric ionic liquid-type GPE was reported by Kuo et al. (2016). The ionic conductivity was 0.12 × 10^{-3} S/cm (at RT) as compared to PVdF-HFP GPE (voltage window = 4.5 V). The initial discharge capacity was 152 mAh/g (0.1 °C) and 117 mAh/g (3 °C) with an efficiency of 99%

(after 100 cycles). The effect of organopolysilazanes (OPSZ) ceramic having different morphologies were investigated on the electrochemical properties of PAN polymer [Smith et al. 2017]. The ionic conductivity was 1.04 ± 0.05 mS/cm for the 40 wt % TEOS:PSZ system and is attributed to large electrolyte uptake due to smaller pore size. The initial discharge capacity of the fabricated cell was 134 mAh/g. However, after 100 cycles the capacity retention is 93% (for 40 wt % TEOS:PSZ), 91% (for 20 wt % TEOS:PSZ). The capacity retention after TEOS:PSZ was enhanced as compared to TEOS:PSZ-free system (88%).

The properties of polymer electrolytes can be tuned by the copolymerization and blending technique because it enhances the electrolyte uptake and amorphous content. Shi et al. (2018) prepared a GPE (PE-PM-PVH) by blending the PEO and PMMA with P(VDF-HFP), and LiPF$_6$- EC+DMC (1:1 v/v) was used as the plasticizer. The ionic conductivity was 0.81 mS/cm and is higher than pristine P(VDF-HFP) (0.25 mS/cm). The Li$^+$ transport number (t_{Li}^+) was 0.72, and the voltage stability window was close to ~5.0 V [for pristine P(VDF-HFP) is 4.5 V]. The fabricated cell shows the discharge capacity of about 152.7 mAh/g and remains 149.6 mAh/g even after 100 cycles (capacity retention = 98%, Coulombic efficiency = 100%).

Another important strategy that is being focused on research is the tuning of the electrode and electrolyte material. Du et al. (2020) reported the fabrication of SC using poly(3,4-ethylene dioxythiophene)/carbon paper (PEDOT/CP) as an electrode and gel polymer [(1-butyl-3-methyl imidazole tetrafluoroborate)/polyvinyl alcohol/sulfuric acid (IL/PVA/H$_2$SO$_4$)] as an electrolyte. The maximum specific capacitance was 86.81 F/g at 1 mA/cm^2 (capacity retention = 71.61 after 1000 cycles). The energy density was 176.90 Wh/kg, and the power density is 21.27 kW/kg. The strong crosslinking points are generated by the freezing–thawing (F/T) method, and PVA (Polyvinyl alcohol) hydrogels are prepared. The number of F/T cycles is crucial. The specific capacitance increased with the increase of F/T cycles (up to F/T 3) and is 53.73 F/g and then decreases. The increase of F/T cycles increases the number of H-bonds in a polymer gel, and a 3D crosslinking network is formed that allows easier access to ion migration [Wu et al. 2011].

Polyelectrolyte (PE)-based GPE is another efficient electrolyte due to high water retention ability that provides ion migration channels for ions in electrolyte [Wang et al. 2017b]. Therefore, Yan et al. (2020) prepared the PE material by the UV-assisted copolymerization of aprotic monomer N-[(2-methacryloyloxy)ethyl]-N,N-dimethylpropanammonium bromide (C$_3$(Br)DMAEMA) and poly(ethylene glycol) methacrylate (PEGMA). The conductivity of the PGPE was 66.8 S/cm at 25 °C. The CV curve of the SC cell was almost rectangular, and a specific capacitance of 64.92 F/g was observed at 1 A/g and 67.47 F/g at 0.5 A/g. The capacity retention was 84.74 % at 0.5 A/g. The SC cell shows an energy density of 9.34 Wh/kg and a power density of 2.26 kW/kg. The capacity retention was 94.63% after 10,000 cycles at 2 A/g and indicates desirable cyclic stability.

Another important electrolyte category is "redox-active electrolytes" and is prepared by the addition of redox additives like methylene blue (MB), iodide salts (KI and NaI), hydroquinone (HQ), p-diphenylamine, etc. [Senthilkumar et al. 2012]. Yadav et al. (2019) prepared a GPE using 1-butyl-3-methylimidazolium bis (trifluoro-methylsulfonyl)imide (BMITFSI) as IL, and sodium iodide (NaI) as redox

(a)   (b)

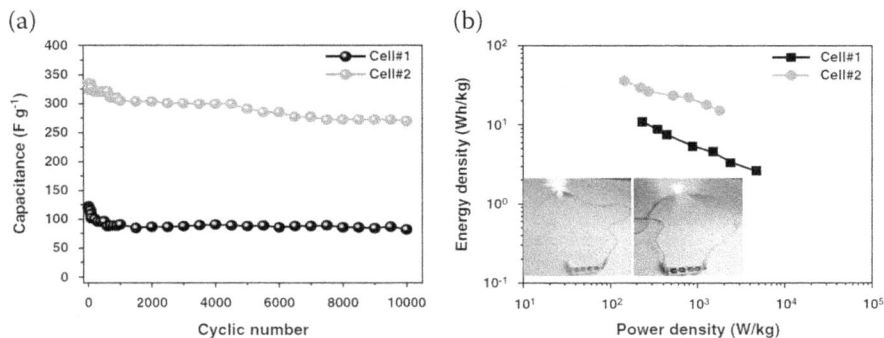

FIGURE 5.6 (a) Specific capacitance of cell 1 and cell 2 versus charge/discharge cycles measured at constant current density 0.84 A/g, and (b) Ragone plots of cell 1 and cell 2 (inset shows glow of LED by four cells connected in series) [Reproduced with permission from Yadav et al., 2019, © Wiley 2019].

additive with poly(vinylidene fluoride-co-hexafluoropropylene) (PVdF-HFP) as host polymer. The specific capacitance was 351 F/g at 5 mV/s for SC cell with redox additive and is higher than redox additive-free cell (128 F/g) [Figure 5.6(a)]. The specific energy was 26.1 Wh/kg, and the power density is 15 kW/kg [Figure 5.6(b)]. Inset shows the LED glow demonstration for 300 s by connecting four cells. The SC cell with redox additive demonstrates superior cyclic stability (5% initial fading) than the redox additive-free SC cell (23% initial fading) for 10,000 cycles.

Another crucial approach to boost the SC performance is by developing composite materials (ion gels) having two networks based on polarity [Peng et al. 2016]. Here, a strongly polar (e.g. PEO, PVA, etc.) network provides superior electrochemical properties, whereas a less polar network (e.g. NBR, natural rubber, PDMS, etc.) leads to enhanced mechanical properties. Based on this, an ion gels composite of PEO/NBR was prepared by in situ synthesis [Lu and Chen 2019]. The ionic conductivity was 2.4 mS/cm for 60% uptake of IL. Then using PEO/NBR ion gels, an SC cell was fabricated using graphene electrodes and tested in the voltage window of 0–2.5 V. The specific capacitance was 280 F/g at 1 A/g and decreases to 150 F/g at 10 A/g. The SC cell demonstrates good cyclic stability up to 10,000 cycles (93.7% capacity retention) and negligible structural degradation as evidenced by XRD (after 10,000 cycles). The energy density was very high 181 Wh/kg (comparable to commercial LIB) with a power density of 5.87 kWh/kg.

5.3.5 SOLID POLYMER ELECTROLYTE

Solid polymer electrolyte (SPE) is an alternative to ionic liquid-based polymer electrolytes and gel polymer electrolytes. The solid polymer electrolyte has better thermal and mechanical stability, and it also facilitates the fabrication of varied geometry cells. The lower ionic conductivity is an issue but can be eliminated by the addition of suitable nanofiller (different morphologies), nanoclay, etc. The surface area of nanofiller plays an important role in the salt dissociation, and the

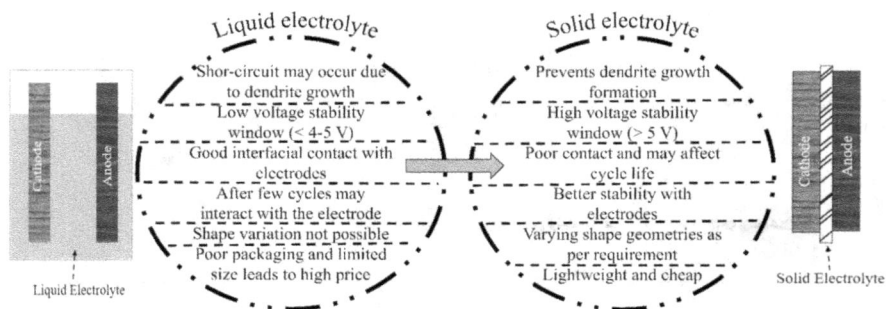

FIGURE 5.7 Benefits of the transition from liquid to the solid electrolyte.

surface group (-OH) provides additional conducting pathways for cation migration. Another attractive strategy to prepare SPE is by selecting two chemically dissimilar polymer segments: (i) aromatic polymer segment and (ii) host polymer matrix. This strategy enables us to develop SPE with better mechanical/thermal stability (by the first segment), and flexibility and low crystallinity (by the second segment) [Arya and Sharma 2019, Arya and Sharma 2020]. Figure 5.7 shows the advantage of a solid electrolyte-based battery.

A solid polymer electrolyte comprises polysulfone, and PEO was prepared with LiTFSI and SN. The highest ionic conductivity was 1.6×10^{-4} S/cm (at RT) and increased to 1.14×10^{-3} S/cm (at 80 °C). The optimum sample depicts a voltage stability window of 4.2 V versus Li/Li$^+$. A full cell was fabricated LiFePO$_4$ (cathode) and Li (anode). The fabricated cell has a discharge capacity of 152 mAh/g (at C/3 rate) and ~125 mAh/g (after 30 cycles). An oxalate-chelated-borate-structure-grafted PVFM-based [poly(vinyl formal)] polymer membrane was prepared by Lian et al. 2014. The voltage stability window was greater than 5 V (versus Li/Li$^+$). The fabricated cell demonstrates a capacity of 137 mAh/g after 20 cycles (Coulombic efficiency was 99.7%).

Another attractive strategy to develop novel SPE is sandwiched structure and has the potential to provide desirable ionic conductivity. One such structure was prepared to comprise PVDF/LLTO-PEO/PVDF [Li et al. 2018]. The sandwiched structure and PVDF layer prevent the chemical reaction of the LLTO and suppress lithium dendrite growth [Figure 5.8(a)]. The ionic conductivity was ~3.01×10^{-3} S/cm [for SWE-III; Figure 5.8(b)], and the voltage stability window was about 5 V (attributed to PEO stability in sandwiched structure). The value of t_+ was 0.70 for sandwiched structure (SWE-III) and is larger than pure PEO(0.54) and PEO(8)+15% LLTO NWs (0.67). This enhancement in the transport number was attributed to the anion trapping ability of LLTO NW. The discharge capacity for cell (LiCoO$_2$, cathode; Li, anode) was 144 mAh/g (at 1 C) and 98 mAh/g (at 5 C). Figure 5.8(c) shows the specific capacity retention plot and is above 91.8% after 100 cycles at a 2 C rate.

A flexible SPE was prepared using 3D nanostructured hydrogel (LLTO) frameworks as a nanofiller, and ionic conductivity was 8.8×10^{-5} S/cm (at 25 °C) [1.5 $\times 10^{-4}$ S/cm at 30 °C] with a voltage stability window of 4.5 V [Bae et al. 2018]. This

(a)

(b)

(c)

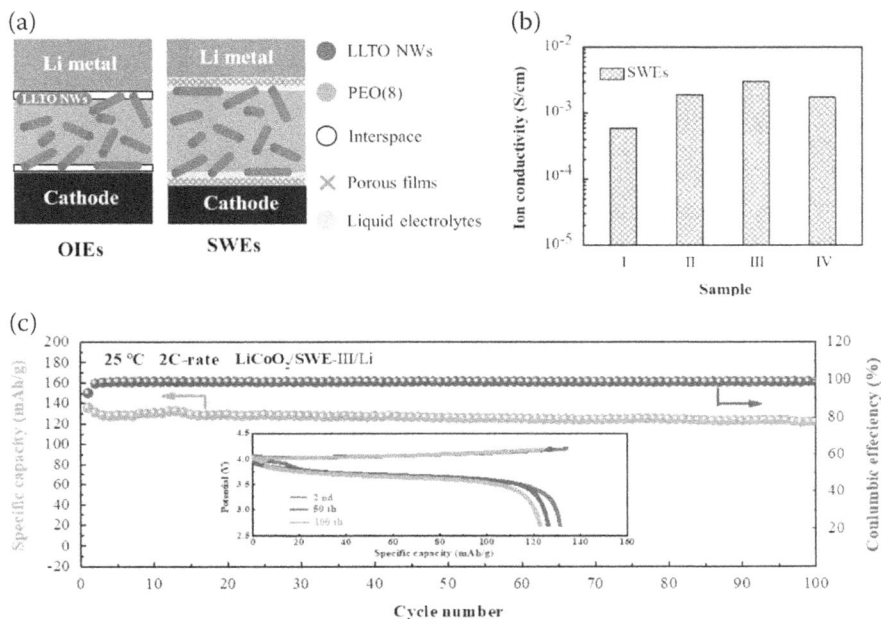

FIGURE 5.8 (a) Sectional diagram of the lithium metal batteries using interlay electrolyte and sandwich structure composite electrolytes, (b) lithium-ion conductivities of the sandwich structure composite electrolytes (SWEs) with a different weight ratio of LLTO NWs. (c) Cycling performance of LiCoO$_2$/SWEs-III/Li battery at 2 C, the insets are the selected charge/discharge curves with different cycles [Reproduced with permission from Li et al. 2018, Elsevier].

increase in conductivity was attributed to the 3D interconnected structure and continuous conducting paths provided by the LLTO framework (hence enhanced ion hopping).

A composite SPE based on the PEO-LiTFSI and garnet Li$_{6.4}$La$_3$Zr$_{1.4}$Ta$_{0.6}$O$_{12}$ (LLZTO) as the nanofiller [Chen et al. 2018]. Figure 5.9(a)–(c) depicts the transition from "ceramic-in-polymer", "intermediate" to "polymer-in-ceramic." The highest ionic conductivity was 1.17×10^{-4} S/cm (at 30 °C) for 10 wt % LLZTO particles [1.58×10^{-3} S/cm at 80 °C]. The enhancement in conductivity was attributed to the faster segmental motion and availability of additional conducting sites for cation provided by LLZTO particles (voltage stability window = 5.0 V versus Li/Li$^+$). The fabricated cell has a discharge capacity of 149.1 mAh/g at 0.1 C (at 55 °C). Figure 5.9(d) shows the cycling stability of cells (LiFePO$_4$ as a cathode and Li as the anode) with electrolyte (PEO-LLZTO-PEG-60 wt % LiTFSI). The discharge capacity was 122.5 mAh/g with Coulombic efficiency of 99.2%. After 50 cycles, capacity increases to 127 mAh/g and Coulombic efficiency was 100%. Figure 5.9(e) displays the LED demonstration with the fabricated cell.

Another composite polymer electrolyte (CPE) comprising PEO and Li$_{1.4}$Al$_{0.4}$Ti$_{1.6}$(PO$_4$)$_3$ (LATP) was prepared and the highest ionic conductivity was

(a)   (b)   (c)

Ceramic–in–polymer   Intermediate   Polymer–in–ceramic

○ LLZTO   ～ PEO   ⊙ LiTSFI   - - - -. Li⁺ pathway

(d)   (e)

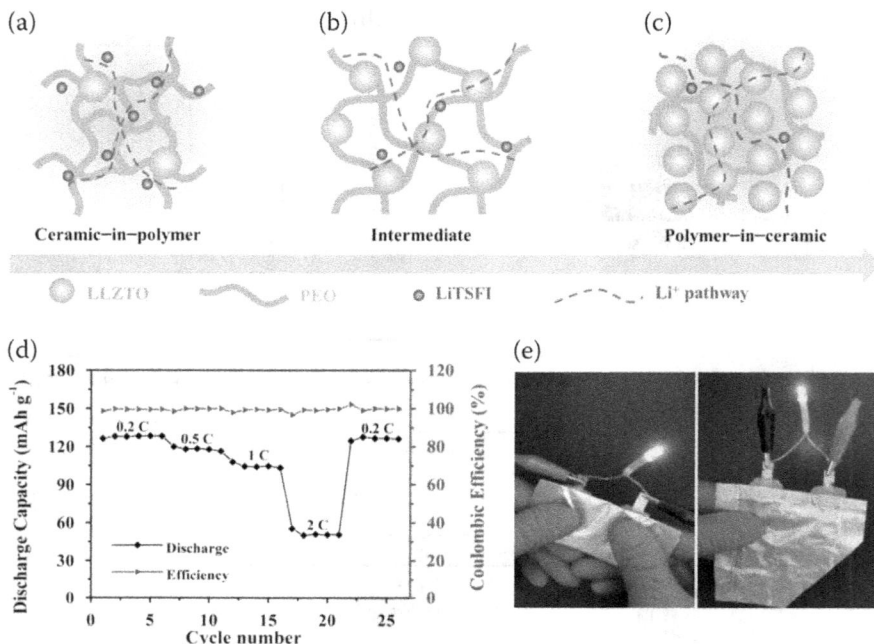

FIGURE 5.9 Schematic illustration for PEO-LLZTO CSE: (a) "ceramic-in-polymer"; (b) "intermediate"; (c) "polymer-in-ceramic"; the typical surface morphologies, (d) rate performance; (e) illustration of solid-state pouch Li metal cell showing well-running under folding and safety with being cut a corner [Reproduced with permission from Chen et al. 2018, Elsevier].

for PEO-LATP01 CPE that is about 6.17×10^{-6} S/cm (at 20 °C) and 7.03×10^{-4} S/cm (at 80 °C) [Liu et al. 2019]. The optimum electrolyte demonstrates a voltage stability window of 4.8 V and enhanced mechanical properties. For the optimum sample, tensile stress increased from 0.35 to 0.95 MPa, and strain decreased from 1572% to 1244%. The fabricated cell ($LiFePO_4$, cathode and Li, anode) shows a capacity value of about 118.3 mAh/g (151.6 mAh/g at 60 °C) and is 89.2% of the theoretical capacity of the cathode.

In SPEs, the dielectric constant of the nanofiller also plays an important role in the enhancement of the ion transport parameters and hence the storage capacity of the SC cell. Therefore, to examine this, Das et al. (2020a) investigated the effect of TiO_2 (dielectric constant: 80) and ZnO (dielectric constant: 8.5) on the polymer matrix of PVDF–HFP incorporating 1-ethyl-3-methylimidazolium tetrafluoroborate ($EMIMBF_4$) as IL. The ionic conductivity of the prepared solid polymer electrolytes was 1.68×10^{-2} S/cm (nanofiller free), 2.57×10^{-2} S/cm (with ZnO), and 3.75×10^{-2} S/cm (with TiO_2) at 303 K. The electrochemical stability window of solid polymer electrolytes was 4.57 V (nanofiller free), 5.55 V (with ZnO), and 5.98 V (with TiO_2). The value of specific capacitance for the SC cell from GCD was 104 F/g (nanofiller free), 131 F/g (with ZnO), and 239 F/g (with TiO_2) at 1 A/g. The increase in specific capacitance for TiO_2 was attributed to a high dielectric constant that supports salt dissociation. The TiO_2–based SC cell shows an energy density of 33.19 Wh/kg and a power density of 1.17 kW/kg. Also, all cells demonstrate a Coulombic efficiency of 100% after 2000 cycles.

Another report from the same group examined the SC performance based on PVdF-HFP polymer matrix, 1-propyl-3-methyl imidazolium bis(trifluoromethyl sulfonyl)-imide as ionic liquid with LiTFSI salt and plasticizer mixture (EC:PC:: 1:1) [Pal and Ghosh, 2018a]. The highest specific capacitance from the CV for cell 3 was 124.1 F/g at a scan rate of 10 mV/s with an energy density of 23.07 Wh/kg and power density of 0.5333 kW/kg. Another report investigated the effect of cationic size and viscosity, the dielectric constant of the ionic liquids on the electrochemical performance of SC cell [Pal and Ghosh, 2018b]. Three polymer electrolytes were prepared: (i) BDMIMBF4-P(VdF-HFP) (*BDMIMGPE–1*), (ii) BMIMBF4-P(VdF-HFP) (*BMIMGPE-2*), and (iii) EMIMBF4-P(VdF-HFP) (*EMIMGPE-3*). The highest ionic conductivity was observed for the EMIMGPE-3 electrolyte (12.76 mS/cm) and is attributed to the smaller cation size, high dielectric constant, and low viscosity. The value of specific capacitance as estimated from CV was 32.66 F/g (cell 1), 49.1 F/g (cell 2), and 63.47 F/g (cell 3) at 10 mV/s. Cell 3 demonstrates 74% capacity retention after 4000 cycles and 100% Coulombic efficiency after 8000 cycles.

Recently, Choi et al. (2019) reported a novel strategy to achieve long-term cyclic stability and high energy density. The authors used the nanofiber cellulose-incorporated nanomesh graphene–carbon nanotube (CNT) hybrid buckypaper electrodes and ionic liquid-based SPE. The SC cell using cPT-200 polymer electrolyte demonstrates areal capacitance of 291 mF cm^{-2} at a current density of 0.75 mA cm^{-2} (capacity retention = 96.3 % after 50000 cycles; at 7.5 mA/cm^2). The effect of bending was examined and capacity retention in bending mode was 98.4% after 50000 cycles. The enhanced performance was attributed to the high ionic conductivity of polymer electrolyte (3.0 mS/cm) and high electrical conductivity of nanofiber cellulose-incorporated nanomesh graphene–CNT hybrid buckypaper (540 S/cm). The SC cell exhibits a gravimetric energy density of 33.6 Wh/kg and volumetric energy density of 6.68 mWh/cm^3.

Recently, Jin et al. (2019) reported the fabrication of SC cells using novel quasi-solid state polymer electrolyte (QPE). The QPE comprises porous acrylate rubber/ tetraethylammonium tetrafluoroborate-acetonitrile (pACM/Et$_4$NBF$_4$–AN) and nitrogen-doped porous graphene (NPG) film-supported vertically aligned polyaniline nanocones (NPG@PANI). The specific capacitance for NPG@PANI-2C electrode cell was 259.5 mF/cm^2 (330.2 F/g; 51.9 F/cm^3) at 1 mA/cm^2. The anode NPG@PDAA-3C demonstrates specific capacitance of about 254.5 mF/cm^2 (294.4 F/g; 50.9 F/cm^2) at 1 mA/cm^2. Then, an asymmetric SC device was fabricated using NPG@PANI as the cathode, NPG@PDAA as the anode, and pACM/ Et$_4$NBF$_4$–AN as polymer electrolyte. The specific capacitance was 6.2 F/cm^3 (124.7 mF/cm^2; 72.1 F/g) at 0.5 mA/cm^2 with capacity retention of 88.7% (after 10,000 cycles). The energy density was 6.18 mWh cm^{-3} (123.5 mWh/cm^2; 71.4 Wh/kg) with power density of 0.033 W/cm^3 (0.668 m/cm^2; 0.386 kW/kg).

To increase the energy density of the SC, various efforts have been done to tune the voltage window of the electrolytes by the addition of IL and plasticizers [Pal et al. 2019]. In continuation to this, Kang et al. (2020) reported the preparation of SPE comprising poly(ethylene glycol) behenyl ether methacrylate-poly[(2-acetoacetoxy)ethyl methacrylate] (PEGBEM-g-PAEMA) graft copolymer by one-pot free-radical polymerization method. The highest ionic conductivity was 1.23×10^{-3} S/cm, and the increase was attributed to high polarity and amorphous nature. Two SC cells were fabricated using PEGBEM-g-PAEMA (cell 1) and PVA/H_3PO_4 (cell 2) as a solid electrolyte and activated carbon as an electrode. Cell 1 demonstrated the specific capacitance of about 55.5 F/g at 1.0 A/g (for cell 2: 40.8 F/g at 1.0 A/g) with a power density of 900 W/kg and energy density of 25 Wh/kg. After bending with an angle of 135°, the performance was better.

5.3.6 HYBRID ELECTROLYTE

The flammable nature of liquid electrolytes and the narrow voltage stability window for polymer electrolytes motivated the researchers to develop hybrid electrolytes (HE) having high conductivity (as of liquid electrolytes) and better safety (like solid electrolytes). This unique configuration has the potential to eliminate the interfacial issues also faced by a solid electrolyte. The HE may also comprise a soft polymer electrolyte (to improve interfacial contact) and rigid inorganic solid electrolyte (to provide high ionic conductivity). Another configuration of HE comprises an inorganic solid electrolyte (provides a path for cation migration) with an outer layer of solid polymer electrolyte to eliminate interfacial challenges. Another configuration is that liquid electrolyte is spread on solid electrolyte to ensure fast ion dynamics and is termed as liquid-oxide hybrid electrolyte [Ye et al 2020].

Kim et al. (2015) prepared the hybrid solid electrolyte (HSE) NASICON ($Na_3Zr_2Si_2PO_1$) for use in a sodium-ion battery. The ionic conductivity of 1.2×10^{-4} S/cm was obtained at 0 °C and increased with temperature to 1.2×10^{-4} S/cm (at 90 °C). The voltage stability window was close to 5 V for HSE and t_+ was 0.92. The fabricated solid-state batteries of half-cells have an initial discharge capacity of 330 mAh/g for a hard carbon anode and 131 mAh/g for a $NaFePO_4$ cathode at a 0.2 C rate. The carbon/HSE/$NaFePO_4$ full pouch cell shows 96% capacity retention after 200 cycles. An aqueous hybrid electrolyte rechargeable battery (AHERB; Zn as an anode, $LiMn_2O_4$ as the cathode) using a neutral-alkaline hybrid electrolyte was fabricated [Yuan et al. 2020]. The average discharge voltage was 0.63 V and is higher than that of an aqueous rechargeable battery at 1.69 C. This battery system delivers a steady energy density of 208 Wh/kg (based on the total weight of active materials) at 1.69 C with a high average output voltage of up to 2.31 V, cycled for over 1000 cycles (Coulombic efficiency of >98%). A hybrid electrolyte (90 wt % inorganic solid and 10 wt % organic liquid) has been prepared by Asl et al. (2012). The inorganic solid electrolyte was $Li_{1.3}Ti_{1.7}Al_{0.3}(PO_4)_3$, and the organic liquid electrolyte selected was 1 M $LiPF_6$ in EC:DEC. The hybrid electrolyte shows bulk resistance of 80 Ω and is much smaller than a cell with a solid electrolyte (420 Ω). The high resistance for solid electrolytes may be due to the presence of cracks and voids that hinder continuous cation migration.

While in the hybrid electrolyte, organic liquid fills this space and provides easy cation migration [Figure 5.10(a)]. Figure 5.10(b) shows the discharge capacity of 100 mAh/g at 0.1 C. A comparison of hybrid electrolyte cells and pure liquid electrolyte cells was made, and hybrid electrolyte cells demonstrated better capacity. Figure 5.10c shows the charge voltage behaviors of both hybrid and liquid electrolyte cells for high temperatures. For a hybrid electrolyte-based cell, voltage drops to 0.9 V from 1.0 V at 80 °C (for liquid electrolyte drops to 0.2 V), and it regains charge up to 1.6 V followed by the cell halt. It confirms the superiority of HE over liquid electrolytes.

Another important type of electrolyte is "water in salt" (WIS) electrolytes and demonstrates better electrochemical stability windows, and its nonflammable nature further strengthens its candidature. But high viscosity and low conductivity hinder its

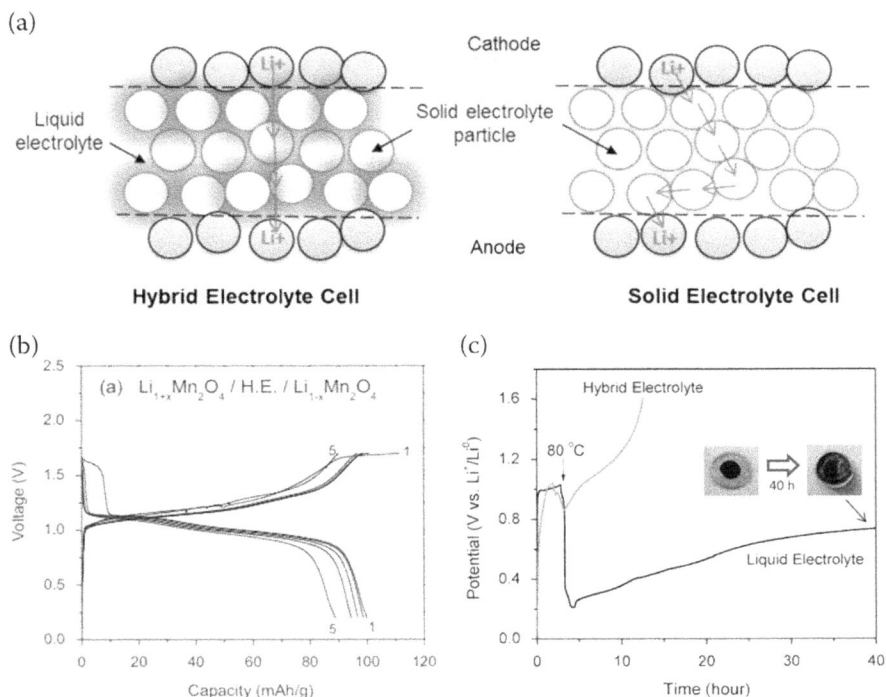

FIGURE 5.10   (a) Schematic diagrams showing the path of Li-ion movement in a solid electrolyte cell and a hybrid electrolyte cell. (b) The first five charge and discharge curves for a hybrid electrolyte (H.E.) cell. (c) Charge voltage curves for a hybrid electrolyte cell and an organic liquid electrolyte during the heat test. As the temperature increases to 80 °C, the voltage of the hybrid electrolyte cell drops from 1.0 to 0.9 V, and then the cell continues to be charged to 1.6 V followed by the cell stops. The liquid electrolyte cell shows a continuous discharge voltage from 0.2 V to below 0.8 V for 40 h after the voltage drops to 0.2 V. The inserted figure shows that the color of the transparent liquid electrolyte changes after reacting with electrode materials at 80 °C [Reproduced with permission from Asl et al. 2012, Elsevier].

use for energy storage devices. A new HE configuration comprising acetonitrile as a co-solvent in WIS electrolyte and is termed as "acetonitrile/water in salt"(AWIS) hybrid electrolyte. This electrolyte provides enhanced conductivity and less viscosity. Dou et al. 2018 prepared AWIS electrolyte, and NMR analysis confirmed the double conductivity (17.4 mS/cm), high ion mobility, and 10% less viscosity for AWIS electrolyte as compared to WIS. The introduction of acetonitrile increases the separation between the lithium ions and TFSI anions, and it was verified from the redshift of the S–N–S bending vibration in Raman spectra (indicating weakened cation–anion interaction). The electrochemical testing of fabricated supercapacitor cells was done at 25 °C. The specific capacitance as obtained from GCD is 27.0 F/g (at 1 A/g) with 81% capacity retention and was higher than WIS electrolyte (22.0 F/g with 42% retention). The full SC cell shows capacity retention of 86% after 6000 cycles (at 1 A/g) and 81% after 14,000 cycles (at 6 A/g) with Coulombic efficiency of 100%.

A garnet-based HSE comprising the PVDF-HFP as a polymer matrix, and LLZO ($Li_7La_3Zr_2O_{12}$) particles were prepared via a tape casting method [Zhang et al. 2018]. The ionic conductivity of the HSE infiltrated with 20 μL liquid

---

TABLE 5.6

Some Available Patents on Electrolytes

Inventor	Polymer used	Patent No.	Year	Conductivity
Bauer et al.	$LiClO_4$ in a 400 MW PEG	4,654,279	1987	4×10^{-4} S/cm at 25 °C
Kuzhikalail M et al.	PAN, EC, PC, and $LiPF_6$/ $LiAsF_6$	5510209	1995	OCV:2.85 V
Nitash Pervez Balsara et al.	Block copolymer	US 8,889,301 B2	2014	1×10^{-4} S cm at 25 °C
Wunder et al.	PEO- POSS-phenyl7(BF_3Li)$_3$	9680182 B2	2017	1×10^{-4} S/cm at 25 °C (for O/Li = 14)
Michael A. Zimmerman	PPS, PPO, PEEK, and PPA	2017/ 0018781 A1	2017	1×10^{-5} S/cm (at RT)
Russell Clayton Pratt et al.	Perfluoropolyether electrolytes terminated with urethane	9923245	2018	3.6×10^{-5} (at 40 °C) 1.1 $\times 10^{-4}$ (at 80 °C)
Mohit Singh et al.	Ceramic electrolyte	20110281173	2018	Stability up to 500 cycles
Nanotek Instruments, Inc.	Hybrid solid-state electrolyte	20180166759	2018	$>10^{-3}$ S/cm, dendrite penetration resistant
Michael A. Zimmerman, Randy Leising	Lithium metal battery with a solid polymer electrolyte	20180151914	2018	1×10^{-3} S/cm at 80 °C/ 1×10^{-5} S/cm at −40 °C
Michael A. Zimmerman, Randy Leising	Solid-state bipolar battery	20180151910	2018	1×10^{-4} S/cm at RT/1 × 10^{-3} S/cm at 80 °C/ cycling efficiency 99%

electrolyte was about 1.1×10^{-4} S/cm (at 25 °C) [7.63×10^{-4} S/cm at 100 °C] and voltage stability window of about 5.3 V (versus Li^+/Li) [for PVDF-HFP, 4.5 V]. The cation transport number (t_+) was 0.61. This increase in electrical properties was attributed to enhanced segmental motion that facilitates faster cation migration via LLZO. The fabricated cell (Li|HSE|LiFePO$_4$) has initial discharge capacity of about 140 mAh/g (at 0.1 C) with Coulombic efficiency of 100 % after 180 cycles (capacity retention = 92.5%).

Table 5.6 summarizes some reported polymer electrolytes, their ionic conductivity, and the electrochemical performance of the cell using them. Table 5.7 shows some patents on the supercapacitor device using different separators. Table 5.8 and Table 5.9 compare the electrochemical performance of different polymer electrolyte-based batteries and supercapacitors, respectively.

TABLE 5.7
Reported Polymer Electrolytes and Fabricated Supercapacitor Performance

Patent application number	Year	Invention
US6356432B1United States	2002	Supercapacitor having a nonaqueous electrolyte and two carbon electrodes.
1263/MUM/2004 A	2006	Polyaniline thin films synthesized by electrochemical anodization at constant potentials. The electrochemical capacitor was formed with H_2SO_4 solution. The specific supercapacitance of 650 F/g and interfacial capacitance of $0.14F/cm^2$ were obtained.
US7226702B2United States	2007	Solid electrolyte made of an interpenetrating network type solid polymer comprised two compatible phases: a crosslinked polymer for mechanical strength and chemical stability, and an ionic conducting phase.
US20100259866A1United States	2010	Fabrication of a supercapacitor by constructing a mat of conducting fibers, binding the mat with an electrolytic resin, and forming a laminate of the electrodes spaced by an insulating spacer.
CN105006377AChina	2015	The composite electrolyte is composed of a blank electrolyte (KOH) and an electrolyte additive (azo substance).
US20170271094A1United States	2016	Polymer supercapacitor fabricated by loading a flexible electrode plate of a high surface area material with metal oxide particles, then encasing the electrode plate in a coating of a polymer electrolyte.
207701 (India)	2017	High-performance electrochemical redox supercapacitors with polyaniline (PANI) as the active material.
US 10, 199, 180 B2	2019	Fabric supercapacitors disclosed herein exhibit great flexibility.

TABLE 5.8

Different Solid Polymer Electrolytes and their Electrical, Transport, Thermal, and Electrochemical Properties

Materials used	Electrical conductivity (S/cm)	Transference number (t^+; cation)	Voltage stability window (V)	Cell configuration	Capacity (mAh/g)	Coulombic efficiency (%)	Capacity retention (%)	References
PVDF/LLTO-PEO/PVDF	~3.01 × 10⁻³	0.67/0.70	5	LCO/SWEs-III/Li	144 (1 C)	–	91.8 (after 100ᵗʰ cycles)	Li et al. 2018
PEO/LLZTO	1.17 × 10⁻⁴ (at 30 °C), 1.58 × 10⁻³ (@ 80 °C)	–	5	LFP/PEO-LLZTO-PEG-60 wt % LiTFSI/Li	139.1 (0.1 C; after 100ᵗʰ cycles)	100 (after 50ᵗʰ cycles)	–	Chen et al. 2018
PEO-LiTFSI/Vertically aligned vermiculite sheets (VAVS)	1.89 × 10⁻⁴ (25 °C)	0.47 (RT)	–	Li/VS-CSPE/LFP	167 (0.1 C)	–	82 (after 200ᵗʰ cycles)	Tang et al. 2019
PEO-LiTFSI-PAGP / PEO/LiBOB/LLZTO	1.6 × 10⁻⁵	– / 0.57	5.0 / ~5.0	LFP/(HSPE or SPE)/Li / Li/PEO/LiBOB/LLZTO / LiFePO₄	100 (0.1 C) / 165.9	>99.5 / 84.9 (100ᵗʰ cycles)	– / –	Piana et al 2019 / Guo et al. 2019
PCPU/PCDL/HDI/DEG/LiTFSI	2.2 × 10⁻⁶ (25 °C), 1.12 × 10⁻⁴ (80 °C)	0.45 (80 °C)	4.5 (80 °C)	LFP/PCPU10-20% Li/Li	128 (0.2 C127 (after 100ᵗʰ cycles)	100	99 (after 100ᵗʰ cycles) 91 (after 600ᵗʰ cycles)	Bao et al. 2018
PEGMA/DLC-((PS)₂₃)d/LiTFSI	1.94 × 10⁻⁴ (30 °C),	0.37	5.1 (30 °C)	Li/SPE/Li	139 (0.1 C), 130 (after 50ᵗʰ cycles) [60 °C]	96, 100 [after 50ᵗʰ cycles] [60 °C]	–	Wang et al. 2018a
P(VDF-HFP)-(PE-PM-PVH)	0.81 × 10⁻³	0.72	~5	–	152.7, 149.6 (after 100ᵗʰ cycles)	99	98	Shi et al. 2018
PEOᴮ2K-POSS	0.16 × 10⁻³ (30 °C) and 0.7 × 10⁻³ (60°C)	–	4.3	Li/PEOB₁₂KPOSS/Li	146.5, 144.5 (after 100ᵗʰ cycles) [0.2 C]	99, 99.7 (after 100ᵗʰ cycles)	–	Fu et al. 2016
PEO/LiTFSI/LLZO	5.5 × 10⁻⁴ (30 °C)	0.207 (60 °C)	5.7	Li° |CPE| LFP	121 (after 100 cycles)	98.9	89 (after 100ᵗʰ cycles)	Chen et al. 2017
PEO-PPC-LiTFSI-LLTO	5.66 × 10⁻⁵ (25 °C), 5.7 × 10⁻⁴ (80 °C)	0.227	5.1	LFP/SPE/Li	135 (0.5 C), 130 (after 100ᵗʰ cycles)	100	96	Zhu et al. 2019
PEO; UHMWPEO-LiClO₄/core-shell protein@TiO₂ NW	1.1 × 10⁻⁴, 2 × 10⁻³ (80 °C)	0.62 & 0.41 (PEO only)	5.4	LCO/CPE/Li (65 °C)	135 (0.2 C)	98.6	94.7 (after 70 cycles)	Fu et al. 2018
PEO-LiTFSI/MXene (Ti₃C₂Tₓ)	2.2 × 10⁻⁵ (28 °C), 0.69 × 10⁻³ (60 °C)	0.18	5.2	LFP|PEO₂₀-LiTFSI-MXene0.02|Li (60 °C)	150 (C/10)	>97 (after 100ᵗʰ)	91.4 (after 100ᵗʰ cycles)	Pan et al. 2019

PEO-LiTFSI/LLTO nanofiber	1.8×10^{-4} (RT)	0.33	4.5	Li|3D-CPE|Li	80 (0.3 C), 25 °C	90–100	–	Wang et al. 2018b
PVDF-HFP/LiTFSI/LLZO nanofibre	9.5×10^{-4} (20 °C)	–	5.2	Li/PVDF-HFP/LiTFSI/LLZO CPE/LFP	140 (0.2 C)	99.9	93 (after 150th cycle 0.5 C)	Li et al. 2019a
PEO-TEGDMA-TEGDME	2.7×10^{-4} (24 °C),	0.56	5.38	Li/PTT-SPE/Li	160 (0.05 C)	–	98.8 (after 100 cycles (0.1 C)	Zhang et al. 2019
PEO-LiClO₄–lepidolite	1.39×10^{-6} (RT), 1.23×10^{-4} (60 °C)	0.72	6	LFP|CPE|Li	120 (0.15 C)	100	–	Wang et al. 2019

*LCO: $LiCoO_2$, LFP: $LiFePO_4$, NCM: $LiNiMnCoO_2$

TABLE 5.9

Reported Polymer Electrolytes and Fabricated Supercapacitor Performance

Polymer electrolyte	Conductivity	Specific capacitance	Capacity retention	Energy density	Power density	References
$(C_3(Br)DMAEMA)$-PEGMA	66.8 S/cm at 25 °C	64.92 F/g at 1 A/g and 67.47 F/g at 0.5 A/g	84.74%	9.34 Wh/kg	2.26 kW/kg	Yan et al. 2020
BMITFSI-NaI-(PVdF-HFP)	–	351 F/g at 5 mV/s	95% (after 10,000 cycles)	26.1 Wh/kg	15 kW/kg	Yadav et al. 2019
$EMIMBF_4$-P(VdF-HFP)	12.76 mS/cm	63.47 F/g at 10 mV/s	74% (after 4000 cycles)	18 Wh/kg	1.2 kW/kg	Pal and Ghosh, 2018b
PVA-H_2SO_4-HQ	29.3 mS/cm	491.3 F/g (0.5 A/g)	82.9%	18.7 Wh/kg	245 W/kg	Zhong et al. 2015
PEO/NBR	2.4 mS/cm	150 F/g at 10 A/g	93.7% (after 10,000 cycles)	181 Wh/kg	5.87 kW/kg	Lu and Chen 2019
PVDF–HFP–$(EMIMBF_4)$–TiO_2	3.75×10^{-2} S/cm (with TiO_2)	206.4 F/g	100%	33.19 Wh/Kg	1.17 kW/kg	Das and Ghosh 2020b
(PVA)/CH_3COONH_4/ BmImCl	7.31 mS/cm	27.76 F/g	–	2.39 Wh/kg	19.79 W/kg	Liew et al. 2014
PVDF-HFP/EMimTFSI þ LiTFS	4.5 mS/cm	108 F/g	–	15 Wh/kg	213 W/kg	Kumar et al. 2012
(poly(VA-co-AN))-1-ethyl-3-methylimidazolium (IL)/ $LiBF_4$	2×10^{-4} S/cm at RT and 7×10^{-3} S/cm at 100 °C	80 F/g (1 A/g)	99% (after 1000 cycles)	61 Wh/kg	500 W/kg	Karaman et al. 2019
pACM/Et4NBF4-AN	–	6.2 F cm^{-3} (124.7mF cm^{-2}; 72.1 F/g) at 0.5 mAcm^{-2}	88.7% (after 10,000 cy.)	6.18 mWh cm^{-3} (123.5 mWh cm^{-2}; 71.4Wh/kg)	0.033Wcm^{-3} (0.668 mWc^{-2}; 0.386 kWkg^{-1})	Jin et al. 2019
PEGBEM-g-PAEMA	1.23×10^{-3} S/cm	55.5 F/g at 1.0 A/g	–	25 Wh/kg	900 kW/kg	Kang et al. 2020
chitosan (CS), starch, glycerol, $LiClO_4$	3.7×10^{-4} S/cm	133 (10 mV/s)	–	50 Wh/kg	8000 W/kg	Sudhakar and Selvakumar 2012

REFERENCES

Allen, J.L., Wolfenstine, J., Rangasamy, E., & Sakamoto, J. 2012. Effect of substitution (Ta, Al, Ga) on the conductivity of $Li_7La_3Zr_2O_{12}$. *Journal of Power Sources*, 206, 315–319. doi:10.1016/j.jpowsour.2012.01.131

Arof, A.K., Amirudin, S., Yusof, S.Z., & Noor, I.M. 2014. A method based on impedance spectroscopy to determine transport properties of polymer electrolytes. *Physical Chemistry Chemical Physics*, 16(5), 1856–1867.

Arya, A., & Sharma, A.L. 2017a. Polymer electrolytes for lithium ion batteries: a critical study. *Ionics*, 23(3), 497–540.

Arya, A., & Sharma, A.L. 2017b. Insights into the use of polyethylene oxide in energy storage/conversion devices: a critical review. *Journal of Physics D: Applied Physics*, 50(44), 443002.

Arya, A., & Sharma, A.L. 2019. Electrolyte for energy storage/conversion (Li+, Na+, Mg 2+) devices based on PVC and their associated polymer: a comprehensive review. *Journal of Solid State Electrochemistry*, 23(4), 997–1059.

Arya, A., & Sharma, A.L. 2020. A glimpse on all-solid-state Li-ion battery (ASSLIB) performance based on novel solid polymer electrolytes: a topical review. *Journal of Materials Science*, 55(15), 6242–6304.

Asl, N.M., Keith, J., Lim, C., Zhu, L., & Kim, Y. 2012. Inorganic solid/organic liquid hybrid electrolyte for use in Li-ion battery. *Electrochimicaacta*, 79, 8–16.

Bachman, J.C., Muy, S., Grimaud, A., Chang, H.H., Pour, N., Lux, S.F., .. & Shao-Horn, Y. 2016. Inorganic solid-state electrolytes for lithium batteries: mechanisms and properties governing ion conduction. *Chemical reviews*, 116(1), 140–162.

Bae, J., Li, Y., Zhang, J., Zhou, X., Zhao, F., Shi, Y., .. & Yu, G. 2018. A 3D nanostructured hydrogel-framework-derived high-performance composite polymer lithium-ion electrolyte. *Angewandte Chemie International Edition*, 57(8), 2096–2100.

Balo, L., Gupta, H., Singh, V.K. & Singh, R.K. 2017. Flexible gel polymer electrolyte based on ionic liquid EMIMTFSI for rechargeable battery application. *Electrochimica Acta*, 230, 123–131.

Bandara, T.M.W.J., Dissanayake, M.A.K.L., Albinsson, I., & Mellander, B.E. 2011. Mobile charge carrier concentration and mobility of a polymer electrolyte containing PEO and Pr4N+ I– using electrical and dielectric measurements. *Solid State Ionics*, 189, 63–68.

Bao, J., Shi, G., Tao, C., Wang, C., Zhu, C., Cheng, L., .. & Chen, C. 2018. Polycarbonate-based polyurethane as a polymer electrolyte matrix for all-solid-state lithium batteries. *Journal of Power Sources*, 389, 84–92.

Bron, P., Johansson, S., Zick, K., Schmedt Auf Der Gunne, J., Dehnen, S., & Roling, B. 2013. Li10SnP2S12: an affordable lithium superionic conductor. *Journal of American Chemical Society*, 135, 15694–15697. doi:10.1021/ja407393

Cao, C., Li, Z.B., Wang, X.L., Zhao, X.B., & Han, W.Q. 2014. Recent advances in inorganic solid electrolytes for lithium batteries. *Frontiers in Energy Research*, 2, 25.

Chavhan, M.P., & Ganguly, S. 2017. Charge transport in activated carbon electrodes: the behaviour of three electrolytes vis-à-vis their specific conductance. *Ionics*, 23(8), 2037–2044.

Chen, F., Yang, D., Zha, W., Zhu, B., Zhang, Y., Li, J., .. & Sadoway, D.R. 2017. Solid polymer electrolytes incorporating cubic Li7La3Zr2O12 for all-solid-state lithium rechargeable batteries. *Electrochimica Acta*, 258, 1106–1114.

Chen, L., Li, Y., Li, S.P., Fan, L.Z., Nan, C.W., & Goodenough, J.B. 2018. PEO/garnet composite electrolytes for solid-state lithium batteries: From "ceramic-in-polymer" to "polymer-in-ceramic". *Nano Energy*, 46, 176–184.

Cheng, H., Zhu, C., Huang, B., Lu, M., & Yang, Y. 2007, Synthesis and electrochemical characterization of PEO-based polymer electrolytes with room temperature ionic liquids. *Electrochimica Acta*, 52, 5789–5.

Choi, J.W., Cheruvally, G., Kim, Y.H., Kim, J.K., Manuel, J., Raghavan, P., Ahn, J.H., Kim, K.W., Ahn, H.J., Choi, D.S., & Song, C.E. 2007. Poly (ethylene oxide)-based polymer electrolyte incorporating room-temperature ionic liquid for lithium batteries *Solid State Ionics*, 178, 1235–1241.

Choi, Y.J., Jung, D.S., Han, J.H., Lee, G.W., Wang, S.E., Kim, Y.H., & Kim, K.B. 2019. Nanofiber cellulose-incorporated nanomesh Graphene–Carbon nanotube buckypaper and ionic liquid-based solid polymer electrolyte for flexible supercapacitors. *Energy Technology*, 7(5), 1900014.

Das, S., & Ghosh, A. 2020a. Symmetric electric double-layer capacitor containing imidazolium ionic liquid-based solid polymer electrolyte: Effect of TiO_2 and ZnO nanoparticles on electrochemical behavior. *Journal of Applied Polymer Science*, 137(22), 48757.

Das, S., & Ghosh, A. 2020b. Symmetric electric double-layer capacitor containing imidazolium ionic liquid-based solid polymer electrolyte: Effect of TiO_2 and ZnO nanoparticles on electrochemical behavior. *Journal of Applied Polymer Science*, 137(22), 48757.

Dou, Q., Lei, S., Wang, D.W., Zhang, Q., Xiao, D., Guo, H., .. & Yan, X. 2018. Safe and high-rate supercapacitors based on an "acetonitrile/water in salt" hybrid electrolyte. *Energy & Environmental Science*, 11(11), 3212–3219.

Du, H., Wu, Z., Xu, Y., Liu, S., & Yang, H. 2020. Poly (3, 4-ethylenedioxythiophene) based solid-state polymer supercapacitor with ionic liquid gel polymer electrolyte. *Polymers*, 12(2), 297.

Dubal, D.P., Ayyad, O., Ruiz, V., & Gomez-Romero, P. 2015. Hybrid energy storage: the merging of battery and supercapacitor chemistries. *Chemical Society Reviews*, 44(7), 1777–1790.

Dumon, A., Huang, M., Shen, Y., & Nan, C.W. 2013. High Li ion conductivity in strontium doped $Li_7La_3Zr_2O_{12}$ garnet. *Solid State Ionics*, 243, 36–41. doi: 10.1016/j.ssi.2013.04.016

Famprikis, T., Canepa, P., Dawson, J.A., Islam, M.S., & Masquelier, C. 2019. Fundamentals of inorganic solid-state electrolytes for batteries. *Nature Materials*, 18(12), 1278–1291.

Feng, X., Ouyang, M., Liu, X., Lu, L., Xia, Y., & He, X. 2018. Thermal runaway mechanism of lithium ion battery for electric vehicles: A review. *Energy Storage Materials*, 10, 246–267.

Feng, X., Zheng, S., Ren, D., He, X., Wang, L., Cui, H., .. & Ouyang, M. 2019. Investigating the thermal runaway mechanisms of lithium-ion batteries based on thermal analysis database. *Applied Energy*, 246, 53–64.

Fu, K.K., Gong, Y., Dai, J., Gong, A., Han, X., Yao, Y., .. & Hu, L. 2016. Flexible, solid-state, ion-conducting membrane with 3D garnet nanofiber networks for lithium batteries. *Proceedings of the National Academy of Sciences*, 113(26), 7094–7099.

Fu, X., Wang, Y., Fan, X., Scudiero, L., & Zhong, W.H. 2018. Core–shell hybrid nanowires with protein enabling fast ion conduction for high-performance composite polymer electrolytes. *Small*, 14(49), 1803564.

Guo, H.L., Sun, H., Jiang, Z.L., Luo, C.S., Gao, M.Y., Wei, M.H., .. & Zhou, H.J. 2019. A new type of composite electrolyte with high performance for room-temperature solid-state lithium battery. *Journal of Materials Science*, 54(6), 4874–4883.

Jana, M., Khanra, P., Murmu, N.C., Samanta, P., Lee, J.H., & Kuila, T. 2014. Covalent surface modification of chemically derived graphene and its application as supercapacitor electrode material. *Physical Chemistry Chemical Physics*, 16(16), 7618–7626.

Jin, McGinn, Y., & McGinn, P.J. 2013. $Li_7La_3Zr_2O_{12}$ electrolyte stability in air and fabrication of a $Li/Li_7La_3Zr_2O_{12}/Cu_{0.1}V_2O_5$ solid-state battery. *Journal of Power Sources*, 239, 326–331. doi: 10.1016/j.jpowsour.2013.03.155

Jin, J., Mu, H., Wang, W., Li, X., Cheng, Q., & Wang, G. 2019. Long-life flexible supercapacitors based on nitrogen-doped porous graphene@ π-conjugated polymer film electrodes and porous quasi-solid-state polymer electrolyte. *Electrochimica Acta*, 317, 250–260

Joost, M., Kim, G.T., Winter, M., & Passerini, S. 2015. Phase stability of Li-ion conductive, ternary solid polymer electrolytes. *Electrochimica Acta*, 113, 181–185.

Jung, N., Kwon, S., Lee, D., Yoon, D.M., Park, Y.M., Benayad, A., .. & Park, J.S. 2013. Synthesis of chemically bonded graphene/carbon nanotube composites and their application in large volumetric capacitance supercapacitors. *Advanced Materials*, 25(47), 6854–6858.

Kamaya, N., Homma, K., Yamakawa, Y., Hirayama, M., Kanno, R., Yonemura, M., .. & Mitsui, A. 2011. A lithium superionic conductor. *Nature Materials*, 10, 682–686. doi:1 0.1038/nmat306

Kang, D.A., Kim, K., Karade, S.S., Kim, H., & Kim, J.H. 2020. High-performance solid-state bendable supercapacitors based on PEGBEM-g-PAEMA graft copolymer electrolyte. *Chemical Engineering Journal*, 384, 123308.

Karaman, B., Çevik, E., & Bozkurt, A. 2019. Novel flexible Li-doped PEO/copolymer electrolytes for supercapacitor application. *Ionics*, 25(4), 1773–1781.

Karuppasamy, K., Rhee, H.W., Reddy, P.A., Gupta, D., Mitu, L., Polu, A.R., Shajan, X.S. 2016. Ionic liquid incorporated nanocomposite polymer electrolytes for rechargeable lithium ion battery: A way to achieve improved electrochemical and interfacial properties. *Journal of Industrial and Engineering Chemistry*, 40, 168–176.

Kim, G.T., Appetecchi, G.B., Alessandrini, F., Passerini, S. 2007. Solvent-free, PYR 1A TFSI ionic liquid-based ternary polymer electrolyte systems: I. Electrochemical characterization. *Journal of Power Sources*, 171, 861–869.

Kim, J.K., Lim, Y.J., Kim, H., Cho, G.B., & Kim, Y. 2015. A hybrid solid electrolyte for flexible solid-state sodium batteries. *Energy & Environmental Science*, 8(12), 3589–3596.

Kim, J., Lee, J.H., Lee, J., Yamauchi, Y., Choi, C.H., & Kim, J.H. 2017. Research Update: Hybrid energy devices combining nanogenerators and energy storage systems for self-charging capability. *APL Materials*, 5(7), 073804.

Kumar, Y., Pandey, G.P., & Hashmi, S.A. 2012. Gel polymer electrolyte based electrical double layer capacitors: comparative study with multiwalled carbon nanotubes and activated carbon electrodes. *The Journal of Physical Chemistry C*, 116(50), 26118–26127.

Kuo, P.L., Tsao, C.H., Hsu, C.H., Chen, S.T., & Hsu, H.M. 2016. A new strategy for preparing oligomeric ionic liquid gel polymer electrolytes for high-performance and nonflammable lithium ion batteries. *Journal of Membrane Science*, 499, 462–469.

Lee, H., Yanilmaz, M., Toprakci, O., Fu, K. and Zhang, X. 2014 A review of recent developments in membrane separators for rechargeable lithium-ion batteries. *Energy & Environmental Science*, 7, 3857–3886.

Li, H., Li, M., Siyal, S.H., Zhu, M., Lan, J.L., Sui, G., .. & Yang, X. 2018. A sandwich structure polymer/polymer-ceramics/polymer gel electrolytes for the safe, stable cycling of lithium metal batteries. *Journal of Membrane Science*, 555, 169–176.

Li, Y., Zhang, W., Dou, Q., Wong, K.W., & Ng, K.M. 2019a. $Li_7La_3Zr_2O_{12}$ ceramic nanofiber-incorporated composite polymer electrolytes for lithium metal batteries. *Journal of Materials Chemistry A*, 7(7), 3391–3398.

Li, Z., Gao, S., Mi, H., Lei, C., Ji, C., Xie, Z., .. & Qiu, J. 2019b. High-energy quasi-solid-state supercapacitors enabled by carbon nanofoam from biowaste and high-voltage inorganic gel electrolyte. *Carbon*, 149, 273–280.

Lian, F., Guan, H.Y., Wen, Y., & Pan, X.R. 2014. Polyvinyl formal based single-ion conductor membranes as polymer electrolytes for lithium ion batteries. *Journal of Membrane Science*, 469, 67–72.

Liew, C.W., Ramesh, S., & Arof, A.K. 2014. Good prospect of ionic liquid based-poly (vinyl alcohol) polymer electrolytes for supercapacitors with excellent electrical, electrochemical and thermal properties. *International Journal of Hydrogen Energy*, 39(6), 2953–2963.

Liu, L., Chu, L., Jiang, B., & Li, M. 2019. Li1. 4Al0. 4Ti1. 6 (PO4) 3 nanoparticle-reinforced solid polymer electrolytes for all-solid-state lithium batteries. *Solid State Ionics*, 331, 89–95.

Lu, C., & Chen, X. 2019. In situ synthesized PEO/NBR composite ionogels for high-performance all-solid-state supercapacitors. *Chemical Communications*, 55(58), 8470–8473.

Ma, X., Song, X., Yu, Z., Li, S., Wang, X., Zhao, L., .. & Gao, J. 2019. S-doping coupled with pore-structure modulation to conducting carbon black: Toward high mass loading electrical double-layer capacitor. *Carbon*, 149, 646–654.

Mauger, A., Julien, C.M., Paolella, A., Armand, M., & Zaghib, K. 2018. A comprehensive review of lithium salts and beyond for rechargeable batteries: Progress and perspectives. *Materials Science and Engineering: R: Reports*, 134, 1–21.

Murugan, R., Thangadurai, V., & Weppner, W. 2007. Fast lithium ion conduction in garnet-type $Li_7La_3Zr_2O_{12}$. *Angewandte Chemie International Edition*, 46, 7778–7781. doi:1 0.1002/anie.200701144

Nithya, V.D., Selvan, R.K., Kalpana, D., Vasylechko, L., & Sanjeeviraja, C. 2013. Synthesis of Bi2WO6 nanoparticles and its electrochemical properties in different electrolytes for pseudocapacitor electrodes. *ElectrochimicaActa*, 109, 720–731.

Oyedotun, K.O., Masikhwa, T.M., Lindberg, S., Matic, A., Johansson, P., & Manyala, N. 2019. Comparison of ionic liquid electrolyte to aqueous electrolytes on carbon nano-fibers supercapacitor electrode derived from oxygen-functionalized graphene. *Chemical Engineering Journal*, 375, 121906. https://doi.org/10.

Pal, B., Yang, S., Ramesh, S., Thangadurai, V., & Jose, R. 2019. Electrolyte selection for supercapacitive devices: a critical review. *Nanoscale Advances*, 1(10), 3807–3835.

Pal, P., & Ghosh, A. 2018a. Highly efficient gel polymer electrolytes for all solid-state electrochemical charge storage devices. *Electrochimica Acta*, 278, 137–148.

Pal, P., & Ghosh, A. 2018b. Solid-state gel polymer electrolytes based on ionic liquids containing imidazolium cations and tetrafluoroborate anions for electrochemical double layer capacitors: Influence of cations size and viscosity of ionic liquids. *Journal of Power Sources*, 406, 128–140.

Pan, Q., Zheng, Y., Kota, S., Huang, W., Wang, S., Qi, H., .. & Li, C.Y. 2019. 2D MXene-containing polymer electrolytes for all-solid-state lithium metal batteries. *Nanoscale Advances*, 1(1), 395–402.

Peng, X., Liu, H., Yin, Q., Wu, J., Chen, P., Zhang, G., .. & Xie, Y. 2016. A zwitterionic gel electrolyte for efficient solid-state supercapacitors. *Nature Communications*, 7, 11782.

Piana, G., Bella, F., Geobaldo, F., Meligrana, G., & Gerbaldi, C. 2019. PEO/LAGP hybrid solid polymer electrolytes for ambient temperature lithium batteries by solvent-free,"one pot" preparation. *Journal of Energy Storage*, 26, 100947.

Pitawala, J., Navarra, M.A., Scrosati, B., Jacobsson, P., & Matic, A. 2014. Structure and properties of Li-ion conducting polymer gel electrolytes based on ionic liquids of the pyrrolidinium cation and the bis (trifluoromethanesulfonyl) imide anion. *Journal of Power Sources*, 245, 830–835.

Polu, A.R. and Rhee, H.W. 2017, Ionic liquid doped PEO-based solid polymer electrolytes for lithium-ion polymer batteries. *International Journal of Hydrogen Energy*, 42 7212–7219.

Senthilkumar, S.T., Selvan, R.K., Ponpandian, N., & Melo, J.S. 2012. Redox additive aqueous polymer gel electrolyte for an electric double layer capacitor. *RSC Advances*, 2(24), 8937–8940.

Sevilla, M., & Fuertes, A.B. 2014. Direct synthesis of highly porous interconnected carbon nanosheets and their application as high-performance supercapacitors. *ACS Nano*, 8(5), 5069–5078.

Shi, J., Yang, Y., & Shao, H. 2018. Co-polymerization and blending based PEO/PMMA/P (VDF-HFP) gel polymer electrolyte for rechargeable lithium metal batteries. *Journal of Membrane Science*, 547, 1–10.

Simonetti, E., Carewska, M., Di Carli, M., Moreno, M., De Francesco, M., & Appetecchi, G.B. 2017. Towards improvement of the electrochemical properties of ionic liquid-containing polyethylene oxide-based electrolytes. *Electrochimica Acta*, 235, 323–331.

Singh, V.K., Chaurasia, S.K., & Singh, R.K. 2016. Development of ionic liquid mediated novel polymer electrolyte membranes for application in Na-ion batteries. *RSC Advances*, 6, 40199–40210.

Smith, S.A., Williams, B.P., & Joo, Y.L. 2017. Effect of polymer and ceramic morphology on the material and electrochemical properties of electrospun PAN/polymer derived ceramic composite nanofiber membranes for lithium ion battery separators. *Journal of Membrane Science*, 526, 315–322.

Sudhakar, Y.N., & Selvakumar, M. 2012. Lithium perchlorate doped plasticized chitosan and starch blend as biodegradable polymer electrolyte for supercapacitors. *Electrochimica Acta*, 78, 398–405.

Tang, W., Tang, S., Guan, X., Zhang, X., Xiang, Q., & Luo, J. 2019. High-performance solid polymer electrolytes filled with vertically aligned 2D materials. *Advanced Functional Materials*, 29(16), 1900648.

Tarascon, J.M., Masquelier, C., Croguennec, L., & Patoux, S. 2017. A promising new prototype of battery. http://www2.cnrs.fr/en/2659.

Tasaki, K., Kanda, K., Nakamura, S., & Ue, M. 2003. Decomposition of $LiPF_6$ and stability of PF 5 in Li-Ion battery electrolytes density functional theory and molecular dynamics studies. *Journal of The Electrochemical Society*, 150, A1628–A1636.

Tey, J.P., Careem, M.A., Yarmo, M.A., & Arof, A.K. 2016. Durian shell-based activated carbon electrode for EDLCs. *Ionics*, 22(7), 1209–1216.

Tsao, C.H., & Kuo, P.L. 2015. Poly (dimethylsiloxane) hybrid gel polymer electrolytes of a porous structure for lithium ion battery. *Journal of Membrane Science*, 489, 36–42.

Väli, R., Laheäär, A., Jänes, A., & Lust, E. 2014. Characteristics of non-aqueous quaternary solvent mixture and Na-salts based supercapacitor electrolytes in a wide temperature range. *Electrochimica Acta*, 121, 294–300.

Wang, B., Tang, M., Wu, Y., Chen, Y., Jiang, C., Zhuo, S., .. & Wang, C. 2019. A 2D layered natural ore as a novel solid-state electrolyte. *ACS Applied Energy Materials*, 2(8), 5909–5916.

Wang, F., Wu, X., Yuan, X., Liu, Z., Zhang, Y., Fu, L., .. & Huang, W. 2017b. Latest advances in supercapacitors: from new electrode materials to novel device designs. *Chemical Society Reviews*, 46(22), 6816–6854.

Wang, R., Li, Q., Cheng, L., Li, H., Wang, B., Zhao, X.S.,& Guo, P. 2014. Electrochemical properties of manganese ferrite-based supercapacitors in aqueous electrolyte: the effect of ionic radius. *Colloids and Surfaces A: Physicochemical and Engineering Aspects*, 457, 94–99.

Wang, S., Wang, A., Yang, C., Gao, R., Liu, X., Chen, J., .. & Zhang, L. 2018a. Six-arm star polymer based on discotic liquid crystal as high performance all-solid-state polymer electrolyte for lithium-ion batteries. *Journal of Power Sources*, 395, 137–147.

Wang, X., Li, Y., Lou, F., Buan, M.E.M., Sheridan, E., & Chen, D. 2017a. Enhancing capacitance of supercapacitor with both organic electrolyte and ionic liquid electrolyte on a biomass-derived carbon. *RSC Advances*, 7(38), 23859–23865.

Wang, X., Zhang, Y., Zhang, X., Liu, T., Lin, Y.H., Li, L., .. & Nan, C.W. 2018b. Lithium-salt-rich PEO/Li0. 3La0. 557TiO3 interpenetrating composite electrolyte with three-dimensional ceramic nano-backbone for all-solid-state lithium-ion batteries. *ACS Applied Materials & Interfaces*, 10(29), 24791–24798.

Wen, J., Yu, Y., & Chen, C. 2012. A review on lithium-ion batteries safety issues: existing problems and possible solutions. *Materials Express*, 2(3), 197–212.

Wu, J., Gong, X.L., Fan, Y.C., & Xia, H.S. 2011. Physically crosslinked poly(vinyl alcohol) hydrogels with magnetic field controlled modulus. *Soft Matter*, 7, 6205–6212.

Wu, X., Song, K., Zhang, X., Hu, N., Li, L., Li, W., .. & Zhang, H. 2019. Safety issues in lithium ion batteries: Materials and cell design. *Frontiers in Energy Research*, 7, 65.

Wu, Z., Xie, Z., Wang, J., Yu, T., Du, X., Wang, Z., .. & Guan, G. 2020. Simultaneously enhancing the thermal stability and electrochemical performance of solid polymer electrolytes by incorporating rod-like Zn_2 (OH) BO_3 particles. *International Journal of Hydrogen Energy*, 45(38), 19601–19610.

Xu, K. 2004. Nonaqueous liquid electrolytes for lithium-based rechargeable batteries. *Chemical Reviews*, 104(10), 4303–4418.

Yadav, N., Yadav, N., Singh, M.K., & Hashmi, S.A. 2019. Nonaqueous, redox-active gel polymer electrolyte for high-performance supercapacitor. *Energy Technology*, 7(9), 1900132.

Yan, C., Jin, M., Pan, X., Ma, L., & Ma, X. 2020. A flexible polyelectrolyte-based gel polymer electrolyte for high-performance all-solid-state supercapacitor application. *RSC Advances*, 10(16), 9299–9308.

Yan, J., Wang, Q., Wei, T., & Fan, Z. 2014. Recent advances in design and fabrication of electrochemical supercapacitors with high energy densities. *Advanced Energy Materials*, 4(4), 1300816.

Yang, Q., Zhang, Z., Sun, X.G., Hu, Y.S., Xing, H., & Dai, S. 2018. Ionic liquids and derived materials for lithium and sodium batteries. *Chemical Society Reviews*, 47(6), 2020–2064.

Yao, Y., Ma, C., Wang, J., Qiao, W., Ling, L., & Long, D. 2015. Rational design of high-surface-area carbon nanotube/microporous carbon core–shell nanocomposites for supercapacitor electrodes. *ACS Applied Materials & Interfaces*, 7(8), 4817–4825.

Ye, W., Wang, H., Ning, J., Zhong, Y., & Hu, Y. 2020. New types of hybrid electrolytes for supercapacitors. *Journal of Energy Chemistry*, 57, 219–232.

Yongxin, A., Xinqun, C., Pengjian, Z., Lixia, L., & Geping, Y. 2012. Improved properties of polymer electrolyte by ionic liquid PP1.3TFSI for secondary lithium ion battery. *Journal of Solid State Electrochemistry*, 16, 383–389.

Yuan, X., Wu, X., Zeng, X.X., Wang, F., Wang, J., Zhu, Y., .. & Duan, X. 2020. A fully aqueous hybrid electrolyte rechargeable battery with high voltage and high energy density. *Advanced Energy Materials*, 10(40), 2001583.

Zhang, W., Nie, J., Li, F., Wang, Z.L., & Sun, C. 2018. A durable and safe solid-state lithium battery with a hybrid electrolyte membrane. *Nano Energy*, 45, 413–419.

Zhang, Y., Lu, W., Cong, L., Liu, J., Sun, L., Mauger, A., .. & Liu, J. 2019. Cross-linking network based on Poly (ethylene oxide): Solid polymer electrolyte for room temperature lithium battery. *Journal of Power Sources*, 420, 63–72.

Zhong, J., Fan, L.Q., Wu, X., Wu, J.H., Liu, G.J., Lin, J.M., .. & Wei, Y.L. 2015. Improved energy density of quasi-solid-state supercapacitors using sandwich-type redox-active gel polymer electrolytes. *Electrochimica Acta*, 166, 150–156.

Zhu, L., Zhu, P., Yao, S., Shen, X., & Tu, F. 2019. High-performance solid PEO/PPC/LLTO-nanowires polymer composite electrolyte for solid-state lithium battery. *International Journal of Energy Research*, 43(9), 4854–4866.

Zhu, Y., Wang, F., Liu, L., Xiao, S., Chang, Z., & Wu, Y. 2013. Composite of a nonwoven fabric with poly (vinylidene fluoride) as a gel membrane of high safety for lithium ion battery. *Energy & Environmental Science*, 6(2), 618–624.

6 Hydroelectric Cells: The Innovative Approach to Produce the Green Electricity

Aarti and Anurag Gaur
Department of Physics, National Institute of Technology,
Kurukshetra 136119, Haryana, India

CONTENTS

DOI: 10.1201/9781003141761-6

6.1   INTRODUCTION

In this contemporary era, energy sources are playing an important role in deciding the world's fortune. Energy is the prime need for all kinds of activities and therefore playing a crucial role in everyone's lives. Energy sources are divided into two parts: non-renewable and renewable energy sources. Renewable energy sources include biomass, solar, wind, hydropower, geothermal, and marine energies etc. It is also known as an alternate source of energy as it is readily available in nature and never runs out. On the other hand, non-renewable energy sources contain coil, oil, nuclear energy, and fossil fuels etc.

6.1.1   ENERGY SOURCES

The continuous energy demand has created a huge discrepancy between the supply of energy and its usage. Currently, most of the non-renewable energy sources such as fossil fuels, coal, and wood are at the diminishing stage. Unfortunately, the whole community around the globe is depending on traditional energy sources as its primary energy source. Out of the total, around 80% of energy is coming from fossil fuels because they are energy-rich and inexpensive to manufacture. One major problem associated with fossil fuels is their limited supply. Along with this, they produce carbon dioxide in the environment and become a major reason for global warming. The emission of carbon dioxide into the environment could turn out to be more dangerous for the existing and coming generation. Thus, it is an urgent need to generate electricity free from venomous and hazardous byproducts at the appropriate cost. The below Figure 6.1 displays the energy generation by distinct sources and technology. This graph delineates the demand for energy by 2040. The traditional energy sources such as coal, oil, gas, and nuclear exhibit a diminishing trend from 2020, whereas the electricity production from the other ones such as wind, hydro, and

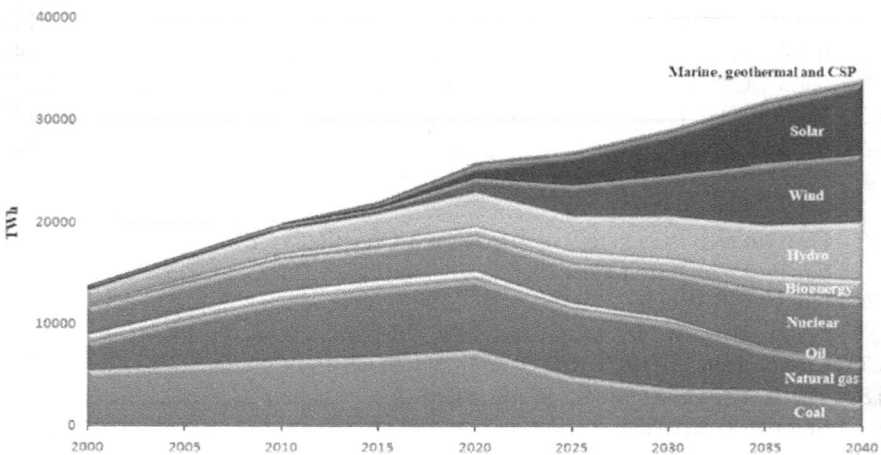

FIGURE 6.1   Electricity production by a variety of energy sources in the globe [https://www.iea.org/reports/world-energy-outlook-2018/electricity].

solar will escalate by 2040, which means that the renewable energy sources would capture more than 50% of total generation. Such a serious situation of high energy demand should not be accepted impulsively rather there could be more stress on energy conservation and exploration of better renewable energy sources.

6.1.2 THE ERA OF GREEN ENERGY

It is also reported by the Intergovernmental Panel on Climate Change that society must shift toward green energy sources that are much better in terms of sustainability, efficiency, and long mean life. Solar energy from the sun is the biggest source of energy on the earth, and it is very difficult to convert and store this solar energy using efficient and economical ways [Pachauri and Meyer 2014]. Researchers have done an ample amount of research in green energy sources, such as organic solar cells, perovskite solar cells, thermoelectric materials, etc. Solar energy availability is limited which depends on the position of the earth. Similarly, wind accessibility is also unpredictable by Mother Nature. For the last decades, researchers have achieved success in energy storage devices in contrast with novel clean energy methods to overcome the sporadic nature of wind and solar generations [Hosseini and Wahid 2020]. Out of all electrical energy storage, the battery is the most convenient, which is adapted in different applications on a large scale, although the battery is a big origin of environmental pollution. Researchers from all over the world are working toward the dissociation of a water molecule by using metal oxides, ferrites, graphene oxide, etc. A graphene oxide energy source can be used to monitor the change in humidity because it produces a high current at a high humidity level. This primary cell in the presence of humidity with an area of 0.1 cm^2 can produce a maximum output power of 2 µW [Wei 2015]. The cement-based batteries deliver current up to 120 µA and open-circuit voltage up to 0.72 V and power output up to 1.4 µW/cm^2 [Meng and Chung 2010]. Recently invented Hydroelectric Cells (HECs) emerging as a promising green energy device along with lots of benefits for the human being. Unlike other batteries like lithium-ion batteries, lead-acid, and solar cells, which are not only toxic but also very expensive, HECs neither produce toxic/greenhouse gases nor use hazardous chemicals. This kind of invention gives us a new route to reduce the use of conventional energy sources and move toward a better option that is safer and environment-friendly. By using only a few drops of water, the HECs devices are able to induce around a quarter ampere current (at less than 1 V), which is more than sufficient to power a small plastic fan. Dr Kotnala has used nanoporous ferrite to split the water molecules in hydroxide and hydronium ions simultaneously. Hydronium ions that are trapped inside the nanopores generate the electric field through which current is produced. Here, zinc and silver are used as electrodes for capturing these ions via capillary and surface diffusion processes [Kotnala and Shah 2016, Kotnala 2018]. The strategy has been built to produce more electricity by choosing a material (ferrites/metal oxides) with high porosity and oxygen deficiency. Along with this, different coping mechanisms and synthesis techniques have been developed to enhance the performance of materials. In the recent time, research is mostly focused on the synthesis of new materials as well as modify the existing ones for the better performance of HECs cells.

6.2   THE BASIC PRINCIPLE OF HECs

In traditional batteries, it is very essential to have electrodes with electrolytes, and it is also essential to have electrolytes in both states either wet state or dry state for chemical reactions to happen. For instance, electrolysis cells require bipolar plates along with a continuous supply of water and electricity on both sides for dissociation of water molecules and having a very complex manufacturing process. In comparison with these conventional cells, HECs are very easy and simple to fabricate which emerged as a suitable candidate with future applications in many areas. A HEC is composed of a nanoporous material pellet, Zn electrode on one face, and Ag electrode on the other side. The material with oxygen-deficient surface cations dissociates the water molecules into hydroxide and hydronium ions at room temperature. Hydroxide ions move toward the Zn anode and oxidize to Zn hydroxide with the formation of two electrons, whereas Ag cathode captures the hydronium ions and these two free electrons, thereby produce H_2 gas and water at the Ag cathode (Figure 6.2).

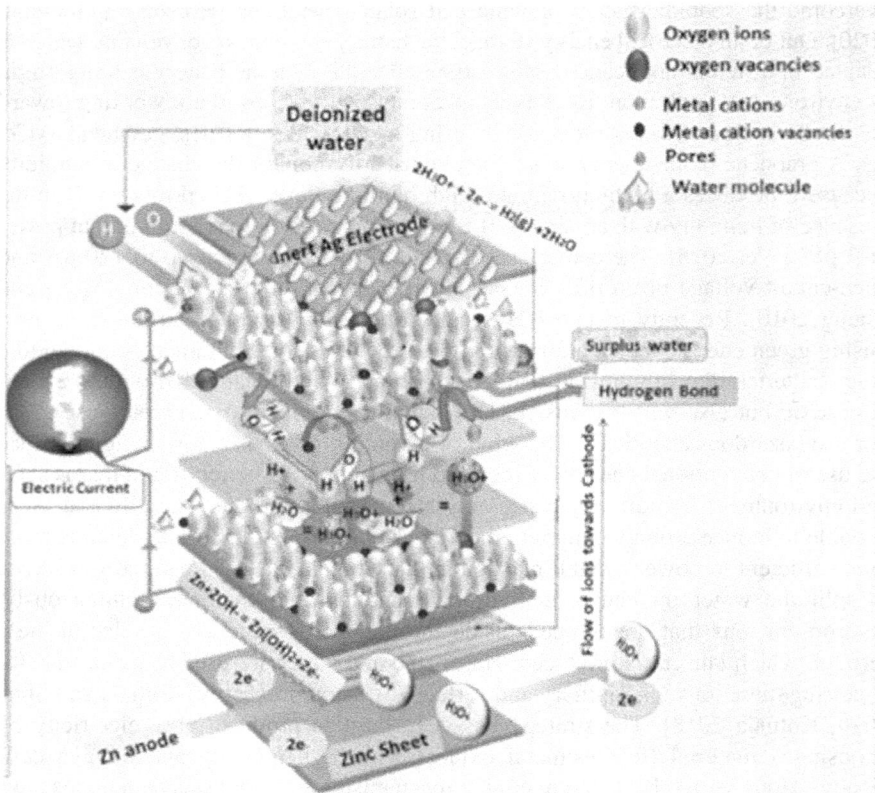

FIGURE 6.2   Basic working principle of HEC [Reproduced with permission from Das et al. 2020].

There are two factors on which the working principle of HECs depend:

- Dissociation of water molecules at room temperature
- Conduction of ions through HECs

Nanopores, unsaturated surface cations, and oxygen defects are the most vital part of a HEC for disunite the polar molecules of water at room temperature. The HEC has two types of surface chemistry: chemidissociation and physidissociation. The adsorbed deionized (DI) water on the HEC surface gets chemidissociated on unsaturated cations present on the surface and the oxygen vacancies trapped the charges because oxygen is highly electronegative in the water molecule. The charged oxygen vacancies that are connected with the oxygen of polar water molecules and surface metal cations undergo a strong Coloumbic interaction that weakens the covalent bond (O-H) of a water molecule. As a result of this, dissociation of water molecules happens over surface cations and vacancies, and OH^- layer forms all over the surface. This chemisorbed OH^- layer gives an enormous amount of surface charge density to further physisorbed the water molecule and release H^+ ions that bounce in the physisorbed water layers. The hydronium ion confines within the pores and produces high electrostatic potential that is enough to dissociate the water molecule [Yamazoe and Shimizu 1986, Von Grotthuß, 1805]. The OH^- and H_3O^+ ions undergo oxidation and reduction mechanisms at the cathode and anode electrodes, respectively, and thus cell voltage is developed. The reaction that occurs on the surface defect of the material is

$$2H_2O = H_3O^+ + OH^-$$

At Zn electrode,

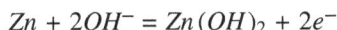

$$Zn + 2OH^- = Zn(OH)_2 + 2e^-$$

At Ag electrode,

$$2H_3O^+ + 2e^- = H_2 + 2H_2O$$

This process of physisorption and chemidissociation of water molecules has been validated by many metal oxide-based HECs.

This ongoing book section briefs the invention of HECs for application in different areas. The selection criteria for the materials will be the focus of the discussion for different types of HECs. The HECs is different from the traditional cells that are of different kinds depending on the material choice and synthesis methods.

6.3 MATERIALS FOR HEC

At present, there is around more than a dozen kind of compound-based HECs have been studied to generate electricity by water splitting. Examples of some compounds are discussed in the following sections.

6.3.1 FERRITE-BASED HECs

The HEC cell, which is first invented by Dr Kotnala, is based on ferrite material. The crystal structure of ferrite is spinel in nature that is extremely stable and existing in the AB_2O_4 form. There is a large number of spinel compounds that have been tried for HECs, but among all of them, magnetite is widely used due to its half-metallic conductivity. Photocatalytic splitting of water molecules on the surface of magnetite electrodes exposes the reaction between water molecules and the magnetite surface [Khader et al. 1987, Ratnasamy and Wagner 2009, Tombácz et al. 2009]. Adsorption of a water molecule on the magnetite thin film proved that the Fe_3O_4 is active toward water molecules [Mulakaluri et al. 2009]. Oxygen vacancies and the defects present on the surface are significant sites for adsorption [Parfitt 1976, Noh et al. 2015]. A nanoporous magnesium ferrite material has been explored to split the water molecules into hydronium and hydroxide ions spontaneously. In comparison to other ferrites, Fe_3O_4 has the highest current due to the presence of maximal surface area. Higher is the surface area, more is the capacity of water adsorption that makes easy ionic conduction, hence higher current.

6.3.2 METAL OXIDE-BASED HECs

HECs are one of the best substitutes for green energy production without any disturbance in the ecological balance of the earth. To improve the research yield in this particular area, metal oxide-based configurations have been used in place of ferrite due to their low-cost availability and versatility. Comparing with ferrite, metal oxide-based HECs providing more electrical power output. Other than this, metal oxide-based materials find vital applications in the field of anti-bacterial agents, solar power generation, electrochemical reaction modulations, and new generation solid-state sensors [Conner and Falcone 1995, Hilmen et al. 1996, Brown et al. 1999, Cortright et al. 2002, Calatayud et al. 2003]. Due to their fascinating electrical and chemical properties and low cost, metal oxide-based materials put forward their promising part in the production of a new class of HECs for green energy synthesis. Metal oxides having a very common point defect on their surfaces known as oxygen vacancies that help to enhance the reactivity of the surface. Many researchers have been studied the dissociation of the water molecule that occurs at defect sites or oxygen vacancies, and this approach is employed for hydrogen production, photocatalytic activities, and other gas sensing applications. Recently, different metal oxides such as SnO_2, TiO_2, Al_2O_3, ZnO, MgO, and SiO_2 have been taken to fabricate HECs and also to verify the interaction of water molecules with oxygen vacancies at room temperature without the use of any alkali/acid and light. The dissociation of water molecules on the surface of these metal oxides and therefore the electric current generation by these metal oxide HECs are shown in Table 6.4. Among all metal oxides, SnO_2 seems to be the best choice for HECs because it has high defect density, low bulk resistance, and low grain boundary resistance that makes it more susceptible to ionic current flow.

6.4 SYNTHESIS AND CHARACTERIZATION TECHNIQUES

There are many approaches to prepare a HEC such as the solid-state reaction method, sol–gel route, and co-precipitation. Recently, these methods are the primary choice for researchers to prepare HECs. In this section, we will discuss some of the synthesis routes for HECs.

6.4.1 SOLID-STATE REACTION METHOD

This route is widely used for the synthesis of polycrystalline solids from a mixture of starting materials that are in high purity form. In this method, the precursor powders are ground by using a pestle and mortar in a proper stoichiometric ratio under an acetone/air environment for 2–4 hours. This fine powder is calcined at different temperature ranges in the presence of air or inert environmental conditions. The obtained powder is again ground for some time before pressing it into the pellet form by using a hydraulic press machine. Pellets are sintered at different temperature ranges to get the required product. R. K. Kotnala et al. have been explored different metal oxides such as SnO_2, Al_2O_3, ZnO, TiO_2, MgO, and SiO_2 based HECs that are prepared using this method [Kotnala et al. 2018]. Lithium-doped magnesium ferrite nanoparticles for the fabrication of HECs of different areas are also prepared by using this route (Figure 6.3).

6.4.2 CO-PRECIPITATION METHOD

Many researchers have used the chemical co-precipitation method to synthesize nanoparticles for the fabrication of HECs. In this technique, the proper molar ratio of precursor salts (such as chloride, nitrate, sulfate) is mixed by using millipore deionized water. S. Jain et al. have used chloride salt of Fe^{2+} and Fe^{3+} in stoichiometric ratio 1:2 to synthesis magnetite nanoparticles [Jain et al. 2018]. The mixture is stirred for several hours by using a magnetic stirrer along with continuous heating. For hematite-based HEC, precursors are heated at 100 ºC for 30 minutes along with stirring [Jain et al. 2018]. Different base solutions like NaOH, NH_4OH, etc. are used for reducing the solution into precipitates. The precipitates are further filtered with acetone and deionized water to neutralize their pH value [Park et al.

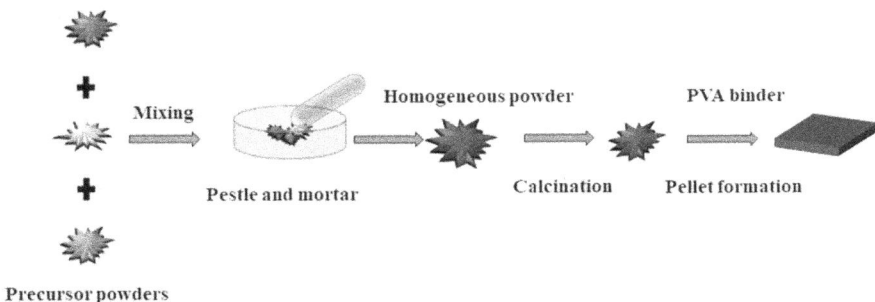

FIGURE 6.3 Schematic representation of solid-state reaction method.

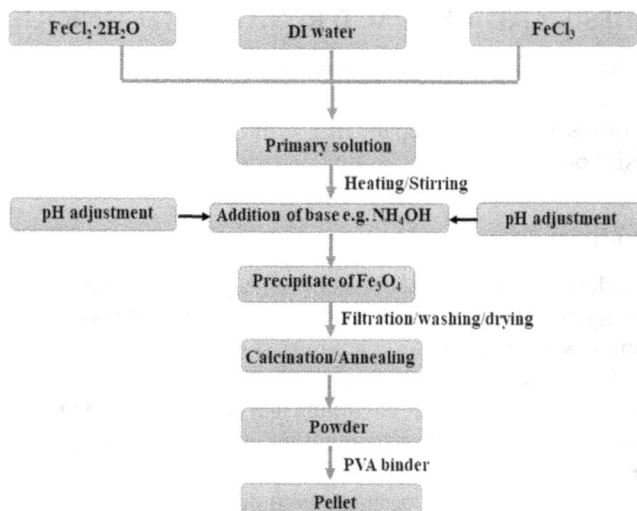

FIGURE 6.4 Co-precipitation route to synthesize magnetite nanoparticles for fabrication of HEC [Reproduced with permission from Jain et al. 2018].

2014, Nawaz et al. 2019]. The final powder undergoes annealing or calcination before pressing into the pellet form (Figure 6.4).

6.4.3 SOL–GEL SYNTHESIS

It is a process that is based on solution chemistry to prepare pure and stoichiometric oxide nanoparticles. A. Gaur et al. prepared Mg-doped and Co-doped SnO_2-based HECs using this method. The precursor's solution of definite volume ratios is mixed under continuous heating and stirring until the formation of a gel. The gel is dried and either annealed or calcined depending on the initial precursor taken. Some benefits are associated with this method, e.g. there is good control on the size of the nanoparticle, chances of getting the predetermined structure of the nanoparticles, and chances to get particles with pure amorphous phases [Khaleel et al. 2013a, Khaleel et al. 2013b, Khaleel and Nawaz 2016. Huang et al. 2019]. The schematic representation of this method is shown in Figure 6.5.

Some illustrations of these techniques and related pros and cons are summing up in Table 6.1 [Reproduced with permission from Nawaz et al. 2019].

6.4.4 FABRICATION OF HECS

A very simple and easy solid-state reaction method is used by the researchers for the initial synthesis of the HEC pellet. First, the material is ground in the pestle and mortar for a particular time depending on the quantity of the material. Pre-sintering and sintering temperature and time adopted by the cell are given in Table 6.2.

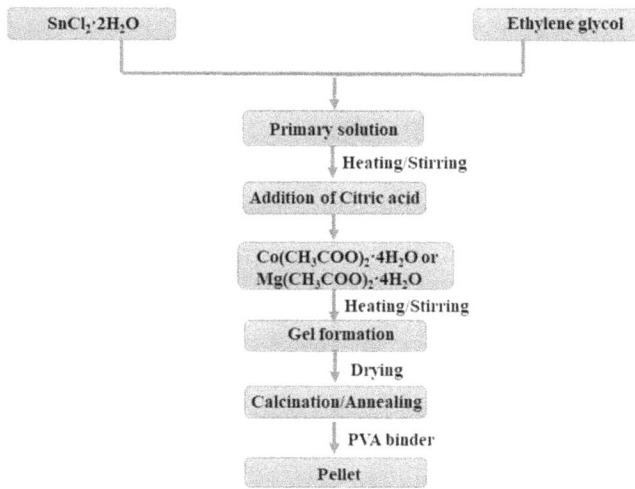

FIGURE 6.5 Schematic representation of the sol–gel synthesis of Co- and Mg-doped SnO$_2$ nanoparticles for the fabrication of HEC.

In general, the pre-sintered powder is again ground for 30 minutes and pressed into a hydraulic press machine for 3–6 ton pressure load to fabricate 2.54 x 2.54 x 0.10 cm^3 pellet. After this, the pellet is sintered and this sintering temperature is optimized according to the material chosen for HEC so that nanoporous microstructure with maximum oxygen vacancies can be produced. Silver paint that is working as a cathode electrode is painted in the comb pattern fashion on one side of the pellet, whereas the Zn plate working as the anode is pasted on the other side. To fully functional HEC, electrical contacts are provided to all the pellets (Figure 6.6 and Table 6.3).

TABLE 6.1
Comparison of Common Techniques to Synthesize Nanoparticles

Synthetic method	Co-precipitation	Sol–gel	Solid-state reaction
Precursors	Nitrates and carbonates	Colloidal solutions	Metal oxides, sulfides, and nitrides
Solvent	DI water	DI water	Not required
Reaction temperature	20–90 °C	Room temperature	500–1500 °C
Morphology	Sphere	Sphere	Sphere
Size	10–50 nm	10–100 nm	≤1 μm
Yield	Scalable	Scalable	High
Advantages	Low-cost chemicals and high yield	Good size control and high yield	Simplicity and large-scale production
Disadvantages	Poor control of size distribution	Low stability in aqueous solution	Very high temperature required

TABLE 6.2

Fabrication Parameters for HECs

Pre-sintering temperature range	600–800 °C
Pre-sintering time	2–8 hours
Sintering temperature range	900–1200 °C
Sintering time	2–4 hours

6.4.5 Characterization Techniques

After the synthesis of nanoparticles, characterization is an essential part. X-ray diffraction, scanning electron microscopy, and transmission electron microscopy are used to analyze the structure and morphology of nanoparticles. Ionic conduction is examined by electrochemical impedance spectroscopy.

6.5 PERFORMANCE PARAMETERS OF HECs on a Qualitative Basis

6.5.1 Material Selection

The selection of material is very necessary while fabricating a HEC. It is important for the material to create oxygen vacancies, porosity, and unsaturated surface cations. Polar molecules form the unsaturated surface cations, and these polar surfaces need to be covered with ion vacancies and hydroxyl groups. For instance, if oxygen ion is removed from TiO_2 by reduction, it exposes two Ti ions of 4-fold coordination. In the existence of anion vacancies, cations with lower coordination form in a lower oxidation state. As a result of this, more sites with oxygen vacancies are created. These types of conditions can also be produced in other oxides such as SiO_2, Cr_2O_3, Al_2O_3,

FIGURE 6.6 The fabricated HECs.

TABLE 6.3
Different Characterization Techniques

Characterization techniques	Parameters extracted
X-ray diffraction (XRD)	• Degree of crystallinity • Crystallite size
Scanning electron microscopy (SEM)	• Microstructure morphology • Elemental composition
High-resolution transmission electron microscopy (HRTEM)	• Particle size • Shape distribution
Brunauer–Emmett–Teller (BET)	• Porosity • Distribution of pores • specific surface area
Electrochemical impedance spectroscopy (EIS)	• Transport parameters (diffusion coefficient, ion mobility, and viscosity)
Keithley 2430 1 kW pulse source meter	• Voltage–current polarization curve

and ZnO [Tsyganenko and Filimonov 1972, Gurlo and Riedel 2007]. To produce nanopores and oxygen vacancies, unique processing steps are used that make the surfaces highly defective/ionic to attract and dissociate water molecules.

6.5.2 Sintering Temperature and Time

The tailored morphology of different HECs is achieved by optimizing the sintering time and temperature. Comparing with the traditional approach, a lower sintering temperature is required for nanoporous morphology. For desired outcomes, the optimal parameters are perceived for getting a morphology with the oxygen-deficient and nanoporous surface. Sintering is one of the vital parameters that influence the pore size, grain size, and morphology significantly.

6.5.3 Porosity

The role of porosity is very important as it is used for the dissociation of water molecules. For the flow of ions, nanopores are very essential. As porosity increases, more water molecules adsorbed on the material surface, hence escalate the dissociation. The factors on which chemidissociation of water molecules depends are electronegativity, nanopores, the number of metal cations, and vacancies formed by oxygen that are present inside and on the surface of the material. Below Figure 6.7 illustrates the porous morphology of various HEC pellets by SEM images. It is visible that the porosity of ferrite-based HEC (Fe_3O_4) is 46% that is very high, whereas least in the case of SnO_2 that is around 6.22%. In the case of Fe_3O_4 and Fe_2O_3, the pore and grain size is minimum; nevertheless, there is a significant rise in the short-circuit current. This rise in current may be due to the presence of unsaturated Fe^{2+} and Fe^{3+} metal cations that are necessary for the dissociation of water

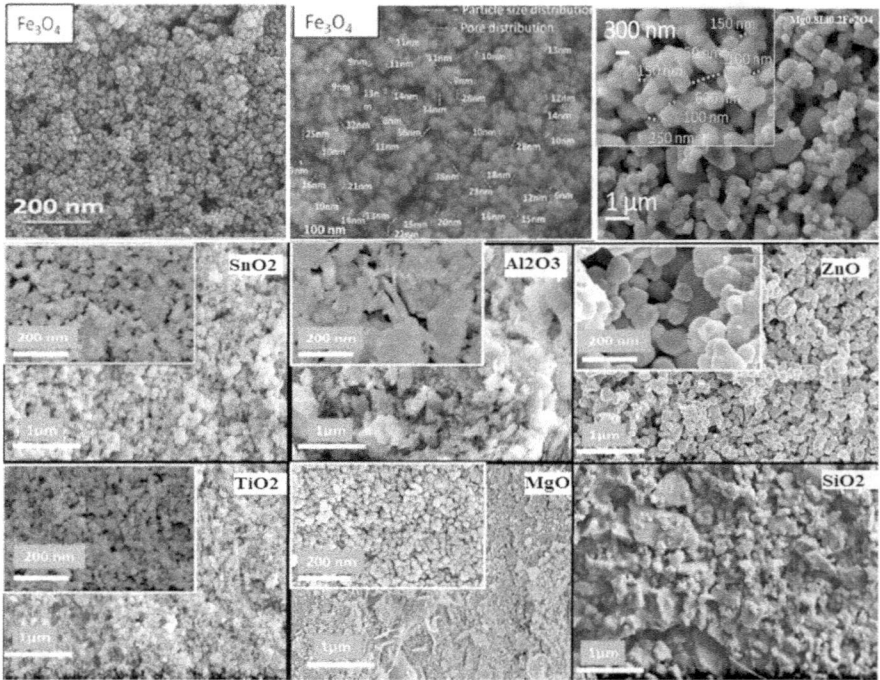

FIGURE 6.7 SEM images of fabricated oxide-based HECs.

molecules. Other metal oxides such as TiO_2, MgO, and ZnO having numerous grain boundaries that obstruct the hopping of H^+ ions to the silver electrode, therefore showing minimum short-circuit current. Similarly, the specific surface area plays a vital role in the adsorption of water molecules. The HECs based on Fe_3O_4, Fe_2O_3, Mg-Li ferrite, and MgO show maximum surface area. As the surface area increases, the capacity of water adsorption increases that results in easy ionic conduction. In the case of MgO, the surface area is more, but the output current is very less. The reason behind this is that the surface current density in MgO is very low that leads to the minimum chemidissociated ions. Similarly, the surface area of SnO_2 and ZnO is very less but gives more current because the chemidissociated ions are very high in the case of SnO_2 and ZnO due to high surface current density. Therefore, for all HECs, porosity is the minimum requirement for the flow of ions (Figure 6.8).

6.6 PERFORMANCE PARAMETERS OF HECS ON A QUANTITATIVE BASIS

6.6.1 POLARIZATION CURVE

A voltage–current polarization curve is used to find out the performance of a HEC. This curve provides information regarding different types of losses such as activation loss, ohmic loss, concentration loss, and internal loss at the surface of the

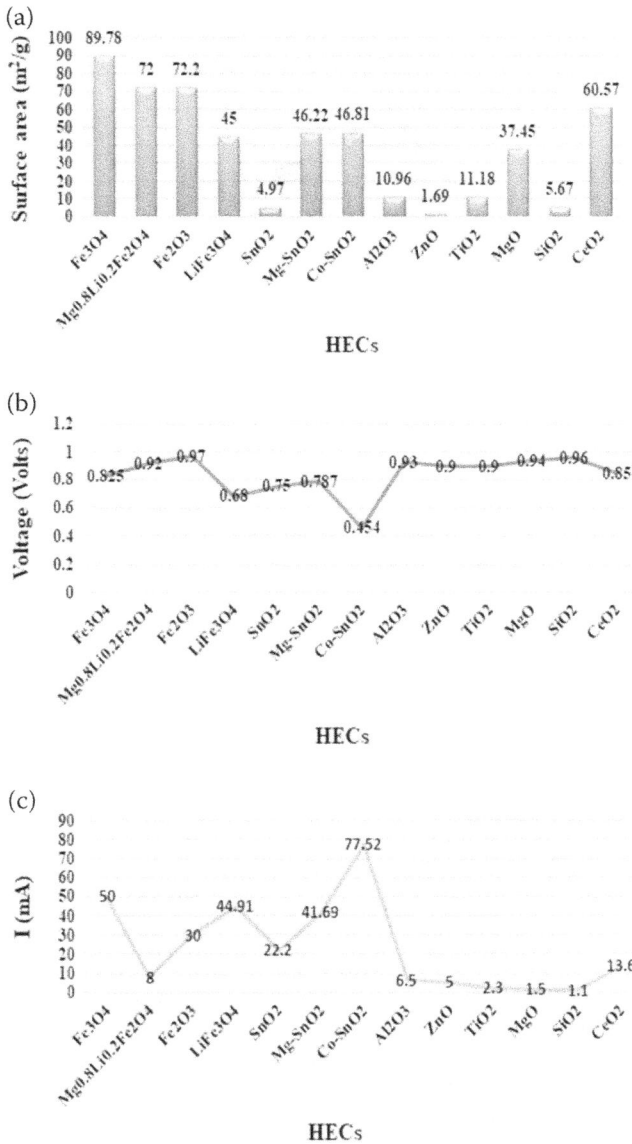

FIGURE 6.8 Graph (a), (b), and (c) delineates the specific surface area, the voltage produced, and current generated by different HECs.

electrode. One may understand the regions of losses within the cell based on this polarization curve. The *V–I* polarization curve for different HECs is shown in below Figure 6.9 [Kotnala and Shah 2016, Kotnala et al. 2018, Jain et al. 2018]. The theoretical value of open-circuit voltage is higher than the measured value due to the presence of these polarization losses. Furthermore, each curve in the graphs is expressing the different regions with different losses. On the surface of a HEC, water

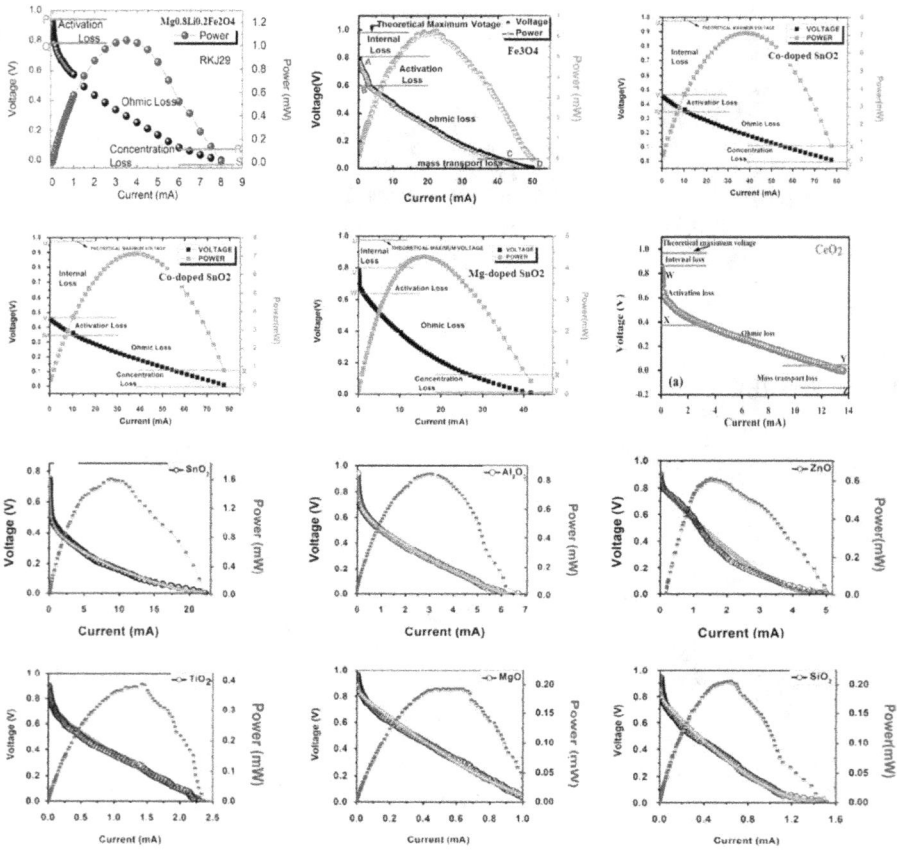

FIGURE 6.9 Polarization curve and power produced by different ferrite- and metal oxide-based HECs soaked in deionized water.

molecules get dissociated into ions and these ions form an energy barrier that interrupts the electrochemical reaction to occur at the material–electrode interface. This is known as activation polarization loss. Ohmic polarization loss is related to the intrinsic resistance of the material. Every material has intrinsic resistance that obstructs the flow of charges. This intrinsic resistance mostly comes from the cell components and comprises interface contacts, grain boundaries, and porosity that affect the voltage and current of the cell. As resistance decreases, voltage decreases and current increases. Lastly, the concentration polarization loss comes due to extreme current densities. The extreme rise in current density is because of the fast utilization of ions by the electrodes that results in a sharp drop in voltage [Jaouen et al. 2002].

6.6.2 Fill Factor

The current–voltage curve represents the correlation between the current flowing through the device and voltage induced or generated across the endpoints. At zero

TABLE 6.4

Fill Factor of Different Metal-Oxide Based HECs

Samples	I_{sc} (mA)	V_{oc} (volts)	Ideal power (mW)	Maximum useful power (mW)	Fill factor	References
Fe_3O_4	50	0.825	41.25	6.5	0.16	[Jain et al. 2018]
$Mg_{0.8}Li_{0.2}Fe_2O_4$	8	0.92	7.36	1	0.14	[Kotnala and Shah 2016]
$Li\text{-}Fe_3O_4$	44.91	0.68	30.80	5.39	0.175	[Gaur et al. 2020b]
SnO_2	22.2	0.75	16.65	1.6	0.10	[Kotnala et al. 2018]
$Mg\text{-}SnO_2$	41.69	0.787	32.81	4.34	0.13	[Gaur et al. 2020a]
$Co\text{-}SnO_2$	77.52	0.454	35.19	7.11	0.20	[Gaur et al. 2020a]
$\alpha\text{-}Fe_2O_3$	30	0.97	27.6	4.5	0.16	[Jain et al. 2019]
Al_2O_3	6.5	0.93	6.04	0.82	0.14	[Kotnala et al. 2018]
ZnO	5	0.90	4.5	0.6	0.13	[Kotnala et al. 2018]
TiO_2	2.3	0.90	2.04	0.4	0.19	[Kotnala et al. 2018]
MgO	1.5	0.94	1.41	0.2	0.14	[Kotnala et al. 2018]
SiO_2	1.1	0.96	1.05	0.2	0.19	[Kotnala et al. 2018]

current, the maximum EMF produced within a HEC is called open-circuit voltage V_{oc}, whereas the highest current flowing through the cell at zero resistance is known as short-circuit current, I_{sc} [Puri 1980, Green 1981, Ramachandran and Menon 1998, Jain and Kapoor 2004].

$$\text{Fill factor} = \frac{I_m \; XV_m}{V_{oc} \; XI_{sc}} = \frac{P_{max}.}{V_{oc} \; XI_{sc}}$$

$I_m \; XV_m$ is the maximum useful power; $V_{oc} \; XI_{sc}$ is the maximum possible power.

One may conclude that the higher the value of the fill factor, the higher is the performance of the HEC (Table 6.4).

6.6.3 Nyquist Plot

Electrochemical impedance spectroscopy (EIS) is widely used for the measurement of ionic current flow through HEC. In this spectroscopy, AC voltage is applied to the device and the response is recorded systematically. In the EIS Nyquist plot, there are three types of semicircle bands depend on the frequency range. Ionic diffusion through the cell is validated by the Nyquist plot that is carried out in wet and dry conditions with a frequency range of 20 Hz–120 MHz and a minimum voltage of 10 mV. Metal oxide-based HECs exhibit very high resistance in a dry condition that can be seen by a semicircular loop in the whole frequency range. To calculate the resistance contribution from electrode, material, and interface individually, an EIS analyzer has been performed when the cell is under wet condition. There are two types of semicircles in the whole frequency range. The first semicircle in the high-frequency region represents the bulk resistance, whereas the second circle that lies in the middle-frequency range is related to the charge transfer resistance. In the case of SiO_2 and MgO, a high-frequency semicircle loop suppresses the Nyquist plot because of the high value of bulk resistance. This high value of bulk resistance prevents the motion of dissociated ions that is confirmed by the EIS fitting value. The value of R_{ct} observed in the Nyquist plot can be associated with the clustering of grains, a large number of grain boundaries, and open macropores. High R_{ct} obstructs the motion of ions at the anode surface that leads to the low value of current. In the case of Al_2O_3, SnO_2, TiO_2, and ZnO, charge transfer resistance is observed that is expressed by a semicircle loop in the middle frequency region. The value of bulk resistance from the Nyquist plot is low in the case of SnO_2, Al_2O_3, and TiO_2 due to low grain boundary impedance that results in the high value of current in these metal oxides, whereas high grain boundary impedance results in high bulk resistance that gives the low value of current in case of ZnO [Kotnala et al. 2018]. In most states, the cells such as Mg-Li ferrite and magnetite exhibit a capacitive tail at a higher frequency region which owing to the production of $Zn(OH)_2$ at Zn anode, whereas another tail at the lower frequency side which explained the diffusion of hydronium ions at silver electrode [Jain et al. 2018]. A. Gaur et. al reported that the value of impedance is of the order of 10^3 ohm in a dry state for Li-doped Fe_3O_4 that declined to 71 ohm in wet conditions, while for Co-SnO_2 and Mg-SnO_2 based HECs the value of impedance is approximately 10^6 ohm for both in dry conditions that further reduces to 54 and 26 ohm in wet states, respectively. Also, there is a tail observed for these cells at the low-frequency side that confirms the diffusion of hydronium ions, hence the high value of current [Gaur et al. 2020a, Gaur et al. 2020b] (Figure 6.10).

6.7 COMPARISON OF HECs WITH TRADITIONAL BATTERIES

A variety of water-based energy sources using external power has been explored to make a comparison of their voltage and current density in the latest invention "HECs." It is found that HECs have an incredible improvement in current density in a short period. Therefore, HECs got potential usage at a low cost (Table 6.5).

6.8 THE LATEST DEVELOPMENT IN THIS FIELD

As we know, spinel ferrites have plenty of applications in the field of high capacity batteries, have magnetic and electric properties, provide gas sensing, cause water splitting, and can be used for wastewater cleaning and photocatalysis. Recently, doping of ferrites and metal oxides has been used by many researchers to enhance the nano-porosity that is a vital part of the physidissociation of water molecules. In this direction, metals like Ni, Co, Mg, Li, Sn, and many more are used as doping elements. The reason behind this is to produce defects within the material. Due to doping, the mismatch between ionic/valence radii of the dopant and the host material creates strain in the lattice that produces oxygen defects. Recently, A. Gaur et al. have fabricated Co-doped SnO_2, Mg-doped SnO_2, and Li-doped Fe_3O_4-based HECs via sol–gel and chemical co-precipitation routes. The pristine SnO_2-based HEC has a specific surface area of 4.96 m^2/g, which escalated to 46.22 and 46.81 m^2/g for Co- and Mg-doped SnO_2 HECs. In the case of Li-doped Fe_3O_4, the specific surface area is 45 m^2/g that is smaller than bare

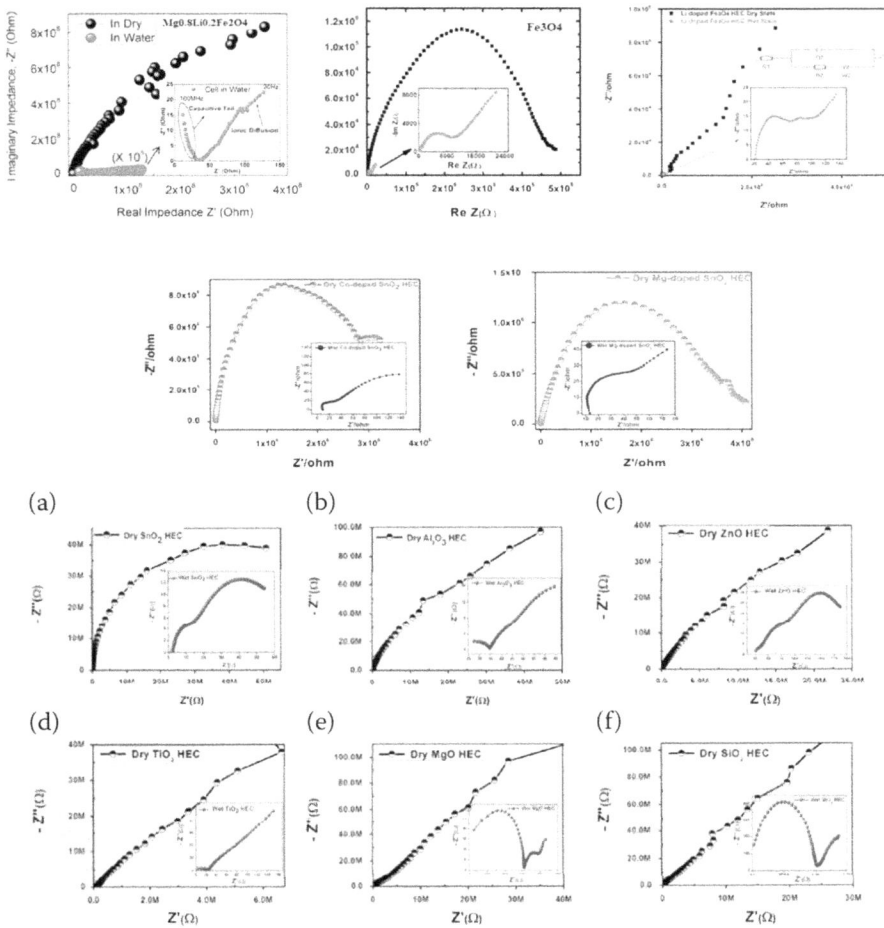

FIGURE 6.10 Nyquist graph of different HECs in both wet and dry states.

Fe_3O_4. More is the surface area, easier is ionic conduction and more is the current obtained [Gaur et al. 2020a,b)]. R. Gupta et al. has fabricated $Al_{2-x}Mg_xO_3$-based HECs for green electricity generation by using a solid-state reaction route. They found that by increasing the Mg doping concentration to $x = 0.5$ in alumina, it will give a maximum current of 15 mA that is more than two order higher than pristine alumina-based HEC [Gupta et al. 2021]. Other than these, work is also going in the field where composites and ceramics have been synthesized to improve the yield. R. Bhargava et al. have synthesized cerium oxide-decorated reduced graphene oxide (CeO_2-rG) nanocomposites to fabricate HEC, whereas J. Shah et al. synthesized multiferroic nanocomposites of BTO and CFO to induce nanoporosity and oxygen defects [Bhargava et al. 2020, Shah et al. 2020]. One of the major applications is demonstrated by burning diyas on Diwali that are made up of HECs. This is not only a cost-effective strategy but also one of the crucial steps to reduce environmental pollution.

TABLE 6.5
Comparison between Various Energy Sources

Items	Lead-acid batteries	Li-ion batteries	Fuel cells	Solar cells	Hydroelectric cells
Inventor	Gaston Plante	John Goodenough	William Robert Grove	Calvin Souther Fuller, Daryl Chapin and Gerald Pearson	R. K. Kotnala and J. Shah
Year of Invention	1859	1980	1839	1954	2016
Resources	Lead, acid electrolyte	Li-ion, anodes: Zn, Al, Cd, Fe etc. Cathodes: MnO_2, HgO, PbO_2, etc.	Costly precursors: Hydrogen and oxygen, carbon paper, etc.	Basically, a p–n junction diode made of silicon	Water, electrodes: Ag, Zn Material: metal oxides/ferrites
Working principle	Convert chemical energy to store potential energy. (When a resistive load is applied across terminals of a battery, the electrical energy can be extracted.)	The electrolyte having Li ions that move from anode to cathode and vice versa through a separator. The battery is charged by a potential difference between the two electrodes	An electrochemical device in which hydrogen reacts with oxygen to produce electricity with heat, and water as by-product, operates at 90 °C	Converts light energy into electrical energy through photovoltaic effect by a p–n junction device	When water is sprinkled on to an HEC's surface, water molecules get chemidissociated into hydronium and hydroxide ions. These ions move toward the respective electrodes resulting in voltage and current in the external circuit
Environmental condition	RT	RT	At or above 90 °C	RT	RT
Electrolyte	Dilute H_2SO_4	LiClO4, $LiBF_4$ suspended in an organic solvent: diethyl and ethylene carbonate	Potassium hydroxide, salt carbonates, and phosphoric acid	No electrolyte is required	No electrolyte is required

By-product	Lead sulfate	Li, Mn, and cobalt oxides	Water	Not useful by-product	Zinc hydroxide, highly pure hydrogen gas (H_2)
Output voltage per cell	2 V	3.6 V	0.70 V	0.60 V	0.45 V
Lifespan	3–5 days	2–3 days	200 days	20 years	2 years
Advantages	Low-cost, safe, durable, and simple to manufacture	Sealed cell, maintenance is not required, long cycle life, high specific and energy density	Clean and efficient at any size, free from greenhouse effect	Only sunlight is required, free from greenhouse effect	No heat is produced during operation, no greenhouse emission, no chemical is used for its working
Disadvantages	Disposal once their chemicals are exhausted	Cost issue, degrades at high temperature, capacity loss when overcharged	High cost safety, lack of fuel and storage	Huge area is required, nonbio degradable storage problem	At the initial stage of prototype energy source is ready for commercialization

Source: Das et al. (2020).

6.9 SUMMARY

HECs are one of the best energy storage devices in this current scenario due to its high current density that is improving day by day. The main parameters are porosity and oxygen defects that are the constraint against its ability. Many researchers are doing more efforts every day to enhance the current density so that it can be an alternative to traditional batteries on a commercial basis. Easy synthesis, lightweight, and various shape forms are attracted researchers to focus more on this area. One challenge in these cells is the production of Zinc hydroxide on the electrode surface that slows down the ionic conduction and makes HECs ineffective after a particular interval of time. We have been discussed different synthesis routes to prepare nanomaterials for the fabrication of a variety of HECs. Furthermore, a comparison between the current–voltage of different HECs has been shown. To conclude, the main goal is to explore more and more materials and ways to improve the performance of HECs to make them commercialized on a large scale. Table 6.6 showing the value of the current–voltage of different HECs that are reported to date.

TABLE 6.6
Comparison between Current and Voltage Generated by All Reported HECs

HECs	I_{sc} (mA)	V_{oc} (volts)	Reference
Fe_3O_4	50	0.825	[Jain et al. 2018]
$Mg_{0.8}Li_{0.2}Fe_2O_4$	8	0.98	[Kotnala and Shah 2016]
$Li\text{-}Fe_3O_4$	44.91	0.68	[Gaur et al. 2020b]
SnO_2	22.2	0.75	[Kotnala et al. 2018]
$Mg\text{-}SnO_2$	41.69	0.787	[Gaur et al. 2020a]
$Co\text{-}SnO_2$	77.52	0.454	[Gaur et al. 2020a]
$\alpha\text{-}Fe_2O_3$	30	0.97	[Jain et al. 2019]
Al_2O_3	6.5	0.93	[Kotnala et al. 2018]
ZnO	5	0.90	[Kotnala et al. 2018]
TiO_2	2.3	0.90	[Kotnala et al. 2018]
MgO	1.5	0.94	[Kotnala et al. 2018]
SiO_2	1.1	0.96	[Kotnala et al. 2018]
$CoFe_2O_4$	0.9	0.95	[Shah et al. 2020]
CeO_2	13.6	0.85	[Bhargava et al. 2020]
$CeO_2.rG1$	16.8	0.835	[Bhargava et al. 2020]
$CeO_2\text{-}rG2$	21.3	0.829	[Bhargava et al. 2020]
$BaTiO_3$	2.0	0.97	[Shah et al. 2020]
$BaTiO_3\text{-}CoFe_2O_4$ (85:15)	7.93	0.70	[Shah et al. 2020]
$Li_{0.3}Ni_{0.4}Fe_{2.3}O_4$	3.8	0.9	[Saini et al. 2020]
$Al_{1.5}Mg_{0.5}O_3$	15	0.9	[Gupta et al. 2021]

REFERENCES

Bhargava R., Shah J., Khan S., and Kotnala R.K. 2020. Hydroelectric cell based on a cerium oxide-decorated reduced graphene oxide (CeO_2-rG) nanocomposite generates green electricity by room-temperature water splitting. *Energy & Fuels*, 34(10), 13067–13078.

Brown G.E., Henrich V.E., Casey W.H., Clark D.L., Eggleston C., Felmy A., Goodman D.W., Gratzel M., Maciel G., and McCarthy M.I. 1999. Metal oxide surfaces and their interactions with aqueous solutions and microbial organisms. *Chemical Review*, 99, 77–174.

Calatayud M., Markovits A., Menetrey M., Mguig B., and Minot C. 2003. Adsorption on perfect and reduced surfaces of metal oxides. *Catalysis Today*, 85, 125–143.

Conner W.C., and Falcone J.L. 1995. Spill over in heterogeneous catalysis. *Chemical Review*, 95, 759–788.

Cortright R.D., Davda R.R., and Dumesic J.A. 2002. Hydrogen from catalytic reforming of biomass-derived hydrocarbons in liquid water. *Nature*, 418, 964–996.

Das R., Shah J., Sharma S., Sharma P.B., and Kotnala R.K. 2020. Electricity generation by splitting of water from hydroelectric cell: An alternative to solar cell and fuel cell. *International Journal of Energy Research*, 44(14), 11111–11134.

Gaur A., Kumar A., Kumar P., Agrawal R., Shah J., and Kotnala R.K. 2020a. Fabrication of a SnO_2-based hydroelectric cell for green energy production. *ACS Omega*, 5(18), 10240–10246.

Gaur A., Kumar P., Kumar A., Shah J., and Kotnala R.K. 2020b. An efficient green energy production by Li-doped Fe_3O_4 hydroelectric cell. *Renewable Energy*, 162, 1952–1957.

Green M.A. 1981. Solar cell fill factors: general graph and empirical expressions. *Solid State Electron*, 24, 788–789.

Gupta R., Shah J., Das R., Saini S., and Kotnala R.K. 2021. Defect-mediated ionic hopping and green electricity generation in $Al_{2-x} Mg_x O_3$ based hydroelectric cell. *Journal of Materials Science*, 56(2), 1600–1611.

Gurlo A., and Riedel R. 2007. In situ and operando spectroscopy for assessing mechanisms of gas sensing. *Angewandte Chemie International Edition*, 46(21), 3826–3848.

Hilmen A.M., Schanke D., and Holmen A. 1996. TPR study of the mechanism of rhenium promotion of alumina-supported cobalt fischer-tropsch catalysts. *Catalysis Letters*, 38, 143–147.

Hosseini S.E., and Wahid M.A. 2020. Hydrogen from solar energy, a clean energy carrier from a sustainable source of energy. *International Journal of Energy Research*, 44(6), 4110–4131. https://www.iea.org/reports/world-energy-outlook-2018/electricity, World energy outlook, n.d.

Huang G., Lu C.H., and Yang H.H. 2019. Magnetic nanomaterials for magnetic bioanalysis, In novel nanomaterials for biomedical. *Environmental and Energy Applications*. Elsevier. 89–109.

Jain A., and Kapoor A. 2004. Exact analytical solutions of the parameters of real solar cells using Lambert W-function. *Sol Energy Mater Sol Cells*, 81, 269–277.

Jain S., Shah J., Dhakate S.R., Gupta G., Sharma C., and Kotnala R.K. 2018. Environment-friendly mesoporous magnetite nanoparticles based hydroelectric cell. *Journal of Physical Chemistry C*, 122, 5908–5916.

Jain S., Shah J., Negi N.S., Sharma C., and Kotnala R.K. 2019. Significance of interface barrier at electrode of hematite hydroelectric cell for generating ecopower by water splitting. *International Journal of Energy Research*, 43(9), 4743–4755.

Jaouen F., Lindbergh G., and Sundholm G. 2002. Investigation of mass-transport limitations in the solid polymer fuel cell cathode: I. mathematical model. *Journal of the Electrochemical Society*, 149(4), A437.

Khader M., Vurens G.H., Kim I.K., Salmeron M., Somorjai G.A. 1987. Photoassisted catalytic dissociation of water to produce hydrogen on partially reduced alpha-iron(III) oxide. *Journal of American Chemical Society*, 109, 3581–3585.

Khaleel A., and Nawaz M. 2016. The effect of composition and gel treatment conditions on the textural properties, reducibility, and catalytic activity of sol-gel prepared Fe (III) {Cr (III) bulk mixed oxides. *Colloids and Surfaces A: Physicochemical and Engineering Aspects*, 488, 52–57.

Khaleel A., Nawaz M., Al-Hadrami S., Greish Y., and Saeed T. 2013a. The effect of metal ion dopants (V^{3+}, Cr^{3+}, Fe^{3+}, Mn^{2+}, Ce^{3+}) and their concentration on the morphology and the texture of doped γ-alumina. *Microporous and Mesoporous Materials*, 168, 7–14.

Khaleel A., Nawaz M., and Hindawi B. 2013b. Sol-gel derived Cr (III) and Cu (II) γ -Al_2O_3 doped solids: Effect of the dopant precursor nature on the structural, textural and morphological properties. *Materials Research Bulletin*, 48(4), 1709–1715.

Kotnala R.K. 2018. Invention of Hydroelectric Cell: A green energy ground breaking revolution. *Journal of Physics Research Application*, 2, 1.s

Kotnala R.K., and Shah J. 2016. Green hydrothermal energy source based on water dissociation by nanoporous ferrite. *International Journal of Energy Research*, 40, 1652–1661.

Kotnala R.K., Gupta R., Shukla A., Jain S., Gaur A., and Shah J. 2018. Metal oxide based hydroelectric cell for electricity generation by water molecule dissociation without electrolyte/acid. *The Journal of Physical Chemistry C*, 122(33), 18841–18849.

Meng Q., and Chung D.D.L. 2010. Battery in the form of a cement-matrix composite. *Cement and Concrete Composites*, 32(10), 829–839.

Mulakaluri N., Pentcheva R., Wieland M., Moritz W., and Scheffler M. 2009. Partial dissociation of water on Fe_3O_4 (001): Adsorbate induced charge and orbital order. *Physical Review Letters*, 103, 176102.

Nawaz M., Sliman Y., Ercan I., Lima-Tenorio M.K., Tenorio-Neto E.T. Kaewsaneha C., and Elaissari A. 2019. Magnetic and pH-responsive magnetic nanocarriers. In *Stimuli Responsive Polymeric Nanocarriers for Drug Delivery Applications*. Woodhead Publishing, 37–85.

Noh J., Osman O.I., Aziz S.G., Winget P. and Bredas J. 2015. Magnetite Fe_3O_4 (111) surfaces: Impact of defects on structure, stability and electronic properties. *Chemistry of Materials*, 27, 5856–5867.

Pachauri R.K., and Meyer L.A. 2014. *Fifth Assessment Report of the Intergovernmental Panel on Climate Change*. IPCC. Geneva. Switzerland.

Parfitt G.D. 1976. Surface chemistry of oxides. *Pure and Applied Chemistry*, 48, 415–418.

Park M., Kim K.Y., Seo H., Cheon Y.E., Koh J.H., Sun H., and Kim T.J. 2014. Practical challenges associated with catalyst development for the commercialization of Li-air batteries. *Journal of Electrochemical Science and Technology*, 5(1), 1–18.

Puri P.S. 1980. Hydrogenation of oils and fats. *Journal of the American Oil Chemists' Society*, 57(11), A850–A854.

Ramachandran R., and Menon R.K. 1998. An overview of industrial uses of hydrogen. *International Journal of Hydrogen Energy*, 23, 593–598.

Ratnasamy C., and Wagner J.P. 2009. Water gas shift catalysis. *Catalysis Reviews Science and Engineering*, 51, 25–440.

Saini S., Shah J., Kotnala R.K., and Yadav K.L. 2020. Nickel substituted oxygen deficient nanoporous lithium ferrite based green energy device hydroelectric cell. *Journal of Alloys and Compounds*, 827, 154334.

Shah J., Verma K.C., Agarwal A., and Kotnala R.K. 2020. Novel application of multiferroic compound for green electricity generation fabricated as hydroelectric cell. *Materials Chemistry and Physics*, 239, 122068.

Tombácz E., Hajdú A., Illés E., László K., Garberoglio G., and Jedlovszky P. 2009. Water in contact with magnetite nanoparticles, as seen from experiments and computer simulations. *Langmuir*, 25(22), 13007–13014.

Tsyganenko A.A., and Filimonov, V.N. 1972. Infrared spectra of surface hydroxyl groups and crystalline structure of oxides. *Spectroscopy Letters*, 5(12), 477–487.

Von Grotthuß T. 2006. Memoir on the decomposition of water and the bodies that it holds in solution with the help of galvanic electricity. *Biochim Biophys Acta*, 1757, 871–875.

Wei D. 2015. Writable electrochemical energy source based on graphene oxide. *Scientific Reports*, 5, 15173.

Yamazoe N., and Shimizu Y. 1986. Humidity sensors: principles and applications. *Sensors and Actuators*, 10(3–4), 379–398.

7 Challenges and Perspectives of Li-Ion Batteries, Supercapacitors, and Hydroelectric Cells

Anil Arya[1], Anurag Gaur[2], and A.L. Sharma[1]
[1]Department of Physics, Central University of Punjab, Bathinda 151401, India
[2]Department of Physics, National Institute of Technology, Kurukshetra 136119, Haryana, India

It may be observed from the detailed discussion (conceptual and experimental) in previous chapters that for the development of next-generation energy storage/ conversion devices (battery, supercapacitor, and hydroelectric cell), new generation materials (electrodes and electrolytes) need to be developed. Because in any device, electrodes and electrolytes play an important role and are directly linked to the device performance. Although existing devices can fulfill the need of energy demand and are also being used in the broad application (from daily use electronic items to space vehicles). However, there exist some shortcomings that restrict the use. Different devices have different challenges and need to be resolved for boosting the energy-based economy globally. A healthy energy economy only can prevent further damage to the environment due to traditional sources of energy. Along with this, these traditional sources are limited; therefore, it becomes important to search for alternate sources of energy for fulfilling the need for energy globally. Therefore, the energy sector can gain momentum if challenges faced by different devices can be resolved. This chapter provides a detailed overview of challenges faced by different devices and their remedies. Although the road to production of next-generation energy storage devices is hard and extensive, the destination is ultimately approached. Figure 7.1 shows the four common challenges faced by different devices. Four important aspects are technical, test standards, economical, and need for smart devices.

Supercapacitors are energy storage devices and have gained the attention of the scientific community due to their high power density, long cycle life, and a broad range of applications. Electrodes and electrolytes are two crucial components and

DOI: 10.1201/9781003141761-7

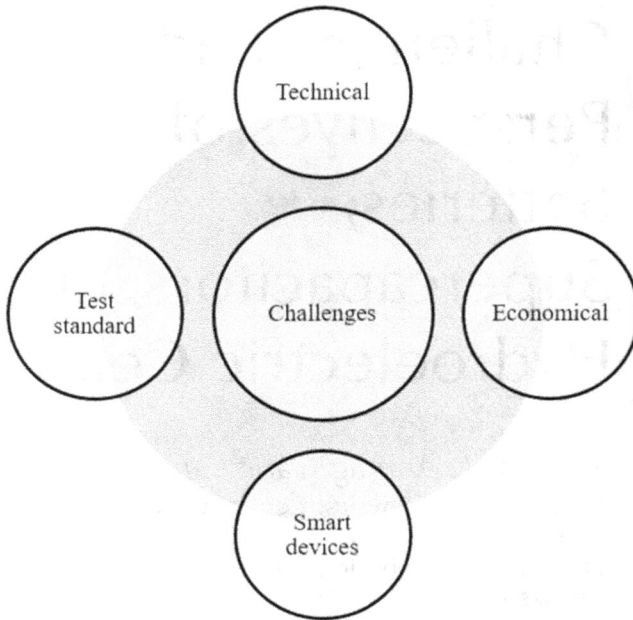

FIGURE 7.1 Challenges for different energy storage/conversion devices.

affect cell performance. Some of the important performance parameters for electrodes are electronic conductivity, pore size, specific surface area, and cost. For electrolytes, ionic conductivity, voltage stability window, thermal/mechanical/ chemical stability, and ion/cation transference number are important parameters. The charge storage mechanism is also different in different systems, and the hybrid system is very complex. Charge storage mechanisms need to be understood in depth for better selection of material as well as to check their potential for safe operation with better cycle life. Still, there is room to improve these parameters but the challenge also on the road. Figure 7.2 depicts the major challenges and is followed by a detailed discussion as well as remedies.

i. **Optimizing the dimension and shape of nanostructures:** As the carbon and oxide materials are key material for supercapacitor cell. Also, their composite demonstrates high performance. The shape and morphology of the electrode material influence the overall electrode nanostructure architecture.
ii. **Optimizing the porosity of nanostructures:** A high surface area is a critical requirement for the electrode material. The high surface area provides more sites for charge storage and along with this porosity also supports the charge storage. The number of pores and the size of the pore are also affecting ion dynamics.
iii. **Enhancing the electronic conductivity of nanostructures:** Electronic conductivity is a crucial parameter as it plays a decisive role in the rate

capability of the supercapacitor cell. The electronic conductivity is lower for the metal oxide-based electrodes as compared to carbon materials. Therefore, by optimising the composition for composite formation the electronic conductivity can be tuned as per requirement.

iv. **Emerging hybrid supercapacitor architecture:** The existing carbon/ metal oxide-based electrode symmetric cell has a lower energy density. Energy density is an important parameter that needs to be improved with the existing high power density. The only feasible strategy is the development of hybrid architecture. In this architecture, the cathode is based on a metal-oxide electrode, and the anode is a carbon electrode. This architecture provides an enhanced voltage window ($E = 0.5CV^2$) of the cell along with enhanced energy/power density. Also, hybrid battery-supercapacitor based on Li/Na/K ion can be developed as they have high power ($\sim$30 kW kg^{-1}) of supercapacitors and a high energy density ($\sim$200 Wh kg^{-1}) of secondary batteries.

v. **Understanding the charge storage mechanism:** The charge storage mechanism is important for exploring the electrode material. As it will allow the tuning of material as per need. The carbon materials store charge via EDLC (Electric Double-Layer Capacitor) mechanism while metal oxides electrodes via the redox reactions. As hybrid architecture is superior as compared to symmetric cells; therefore, the charge storage mechanism needs to be explored for hybrid cells for a better understanding of the performance.

vi. **Need to search for sustainable/green electrode materials:** Energy need is increasing worldwide and some other challenges linked to pollution and cost need to be taken care of. The cost, as well as demand for raw material for electrodes, is increasing; therefore, it needs to be managed for a long-term solution. The raw material needs to be cleaned, and the manufacturing process also contributes to pollution as well as cost and altering environmental conditions. There is a lot of work that needs to be done for exploring the possibility to develop "green/sustainable materials." One unique feature of the green material is its nontoxic nature, large abundance in nature, and can be recycled swiftly.

vii. **Elimination of interfacial and poor stability issues:** The cell performance is strongly influenced by the contact between different components of the cell. The contact between the current collector as well as the active material needs to be improved for minimizing the interfacial resistance and for improving the charge transferability of the electrode material. Also, it facilitates the full use of the active material for redox reactions.

The stability of the SC cell is being influenced by the electrode material also. The electrolyte also affects the stability of the cell; therefore, it needs to be optimized for achieving the long-term cyclic stability of the cell. The main reason for poor stability is the collapse or distortion of the electrode because it reacts with the electrolyte. Therefore, the stability of the electrode and electrolyte material needs to be examined. These stability issues can be eliminated by the selection and optimization of electrode material with electrolytes.

viii. **Launching the supercapacitor cell testing standards:** For commercialization of the supercapacitor cells across the globe, the performance data must of standard. For this, universal testing standards need to be established for better comparison of the performance data from different research groups as well as energy sector industries. It can be seen from the existing literature that most researchers display electrical data for a single electrode only not for the full SC cell. Also, they do energy density and power density calculations for the single electrode only (via testing with a three-electrode cell). These practices can be stopped if there is a proper testing standard available. Also, a lot of literature is available that is reporting charge storage capacitance based on the GCD (Galvanostatic Charge-Discharge) curve in F/g (specific capacitance). However, the material is of battery type and is being confirmed by the nature of the GCD curve. For battery type, a plateau is observed. Therefore, for battery-type materials, the specific capacity needs to be estimated in the unit's mAh/g instead of specific capacitance. For practical application, the GCD estimation is important and needs to be calculated accurately.

ix. **Flexible device and miniaturization**: One last and important challenge that needs to be resolved is the increasing demand for flexible electronic devices and miniaturization. The existing traditional devices restrict the use of these in different application range due to their rigid nature. Therefore, flexible devices with high electrochemical performance will act as a pillar for the future electronic market based on next-generation devices.

x. **Need for Intelligent devices**: As over the globe, the utility of smart and intelligent electronic devices has gained speed. The traditional energy devices are dull as they just store and release energy. But they can't optimize their operation during charging/discharging for achieving long cycle life. They are unable to detect in advance any operation failure or fire. Therefore, by combining AI and deep learning, these devices can be programmed by feeding data. A lot of work needs to be done for the exploration of intelligent devices to eliminate the issues in the energy sector. Some of the important advantages of intelligent devices are large reliability, faster, and maybe developed at a large scale in a small time.

From the detailed discussion on electrolytes in Chapter 5, it is clear that for the development of next-generation energy storage/conversion devices (battery and supercapacitor), new generation electrolytes need to be developed. The performance of devices is linked with the electrolyte because it acts as a carpet for ions. Therefore, a lot of efforts have been already made to improve the electrical, thermal, mechanical, and electrical properties (ionic conductivity, voltage stability window, ion/cation transference number, and ion mobility). Overall, the three important parameters that need to be targeted are conductivity, interfacial stability, and mechanical stability. Therefore, with the careful and controlled tuning of these parameters by engineering, the electrolyte material as well as architecture, the

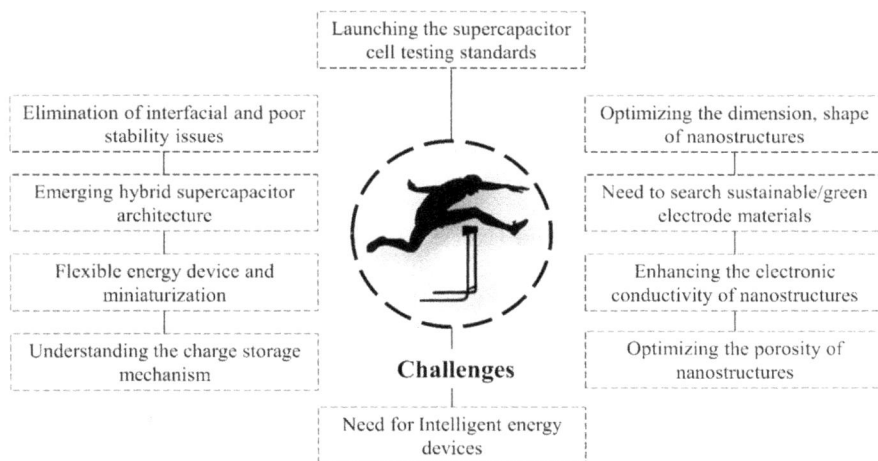

FIGURE 7.2   Challenges for supercapacitor electrode.

dream of a suitable electrolyte can be accomplished. Along with this, the main target is to develop and commercialize all-solid-state batteries and supercapacitors, but still, they are facing many challenges that need to be eliminated before going ahead.

  i. **Moderate electrical properties:** One of the key obstacles for the application of polymer electrolytes is the low ionic conductivity at room temperature. Although it is good in gel polymer electrolytes, for solid polymer electrolytes, the global electrolyte research community is deficient in this. Different strategies (blending, copolymerization, and cross-linking) need to be adopted to engineer the ion dynamic within the polymer electrolytes to realize the high ionic conductivity ($\sim10-3$ S/cm). Some novel composite and solid-state polymer electrolytes have been explored with good ionic conductivity with good mechanical stability. By considering the practical approach, the ionic liquid may be blended with a polymer matrix such that a good balance between conductivity and safety can be attained. Another important parameter is the Li+ transference number, and the high value of this suppresses the concentration polarization and favors fast cation migration. Still, the value obtained is about 0.5. This needs to be improved by tuning the polymer architecture (for faster cation migration via segmental motion) as well by modification of anions for suppressing the anion migration (or trapping the anion with functional groups) with free volume available in the electrolyte. Both the high value of ionic conductivity and the cation transference number will enhance the overall ion mobility.

 ii. **Suppression of dendrites:** Dendrites are threatening the application range of the battery and affect the lifespan of the battery sometimes very dangerous to such an extent that it may damage the battery. For developing the high energy density batteries and supercapacitor, dendrites growth need to be suppressed for long-term safe operation. A lot of effort has already been done by developing nonporous electrolytes and solid polymer electrolytes. Ceramic/sandwiched polymer electrolytes and solid-state electrolytes (e.g. Garnet, NASCION, etc.) are effective in suppressing the dendrites. There is a need to explore electrolyte engineering as well as anode need to be modified structurally for better results.

iii. **Elimination of interfacial issues:** Electrode/electrolyte interface also affects the device performance as well as safe operation. Therefore, this interface needs to be regulated for faster chemical reactions. Electrolyte additives can be used to minimize interfacial resistances. New characterization techniques (e.g. Cryo-electron microscopy) need to be adopted for a better understanding of the interfacial issues.

iv. **Exploration of self-healing polymer electrolytes (SHPE):** Polymer electrolytes are good for energy devices and demonstrate better performance as compared to traditional liquid electrolytes. One unique feature is the flexibility that is an important requirement for the fabrication of foldable and flexible electronic gadgets. But one issue is that this electrolyte cannot be recovered or healed after damage due to internal/external fault. Therefore, polymer electrolytes having the ability to self-repairing and which can restore the ionic conductivity need to be explored. Also, there is a need to develop recyclable electrolyte materials and recycling can be performed without using any severe chemical and high temperature.

v. **Green/sustainable electrolytes:** As the electrolyte is the heart of the battery and is sandwiched between electrodes. The electrolyte is a crucial component of the battery system. Therefore, attention must be devoted to the search for green/sustainable electrolyte materials. The use of green electrolytes will promote a healthy economy. Novel electrolytes need to be explored that affect marginally the environment during production. Along with this, another alternative is the use of biopolymers that can be obtained from the bio-mass or any other secondary raw material.

vi. **Advanced characterization technologies:** Ion dynamics via the segmental motion in polymer electrolytes is crucial to understand the actual effect of additives and their role in ion dynamics. For that existing techniques provide good information, but for in-depth understanding, new techniques need to be adopted. Advanced techniques such as XPS, SSNMR, AFM, cryo-electron microscopy, etc. need to be performed to examine the ion transport parameters as well as complex reactions at the electrode/electrolyte interface. Solid-state nuclear magnetic resonance (SSNMR) may be used to examine the local chemical environments and ion transport conducting pathways of polymer electrolytes.

vii. **Optimization of battery design structure:** Fabrication technique and processing conditions of the device also influence the performance. New battery manufacturing technologies need to be explored for large-scale fabrication. All-solid-state batteries (ASSBs) architecture may be revolutionary for next-generation applications due to long cycle life, high energy density, and particularly better safety. Although ASSB is still a dream for attaining commercialization globally. But it can be achieved by simultaneously focusing on the complete process, i.e. from material selection (for electrode and electrolyte) to full device fabrication.

viii. **Affordable cost:** Cost is a crucial factor to commercialize any product for its global use. Therefore, techniques for the preparation of low-cost material need to be explored. Also, raw materials cost need to be regulated worldwide for uniform use. Some important factors that decide the overall cost of the device are electrode material, separator (not required for ASSBs), electrolyte raw material, packaging, supply chain, and energy economy.

The energy production through hydroelectric cells (HECs), based on metal oxides and their composites, by the splitting of water has turned out to be a feasible alternative to other green energy sources. This way of power generation is purely eco-friendly because the by-products released here are completely nontoxic. The multiple advantages and future prospects of HECs are briefly addressed below.

- HECs use commonly available water as a fuel, which replaces the toxic chemicals that threaten the safety of consumers and harmful to the environment.
- HEC is an eco-friendly alternative to solar cells and fuel cells to provide clean and green energy applications in the future.
- Moreover, the HECs are expected to be inexpensive once mass-produced, even less than the cost of the solar panel.
- HEC is the world's first acid and alkali-free cell and portable.
- HEC provides environment-friendly solution of energy generation scarcity for masses living in remote locations without the use of any acid/alkali.
- HEC is the best alternative source of green energy generation along with its biocompatible by-products in the form of H_2 gas and $Zn(OH)_2$ nanoparticles. Zinc hydroxide nanopowder is produced at the anode by consuming zinc electrodes. Zinc hydroxide can easily be converted to zinc oxide nanopowder via the thermal decomposition process. Zinc oxide possesses multifunctional characters and is used in optoelectronics, semiconductors, paints, paper industries, etc.
- HECs can be developed in different shapes (cylindrical, tubular, circular, etc.) depending upon the necessity of application. It can be successfully used as replacements for lead-acid/Li-ion batteries with different magnitudes of current and power output as a primary source, and there is no need for charging unlike in batteries.
- The energy produced by HEC can be utilized in domestic residential applications in decentralized mode at a low cost because conventional usage of electrical energy is associated with a huge expense of electrical transmission and distribution. It can also supply the electrical needs of the household and in the automotive engines as a clean energy source. Production and usage of such cells are not capital intensive, unlike electrical power generation systems.

- These cells are expected to replace batteries and solar cells for megawatt generators in the future to reach higher overall efficiencies.

The challenge in these cells is the production of zinc hydroxide on the electrode surface that slows down the ionic conduction and makes HECs ineffective after a particular interval of time. The other challenge is to further explore more and more materials and ways to improve the output power of HECs to make them commercialized on a large scale.

Index

For Product Safety Concerns and Information please contact our EU
representative GPSR@taylorandfrancis.com
Taylor & Francis Verlag GmbH, Kaufingerstraße 24, 80331 München, Germany

www.ingramcontent.com/pod-product-compliance
Lightning Source LLC
Chambersburg PA
CBHW060443240326
41598CB00087B/3286